Contents

Preface

Welcome to this workbook. You will probably already be aware that this is the latest edition of one of a long established series that has helped many thousands of students of motor vehicle subjects to success with their examinations. Along with its companion volumes in the Thomson Learning, Vehicle Maintenance and Repair Series it comprehensively covers the area of essential knowledge required by NVQ/SVQ and similar courses in that field.

Both students and teachers will be pleased to know that the book cuts down on the time and effort needed to record information by leaving only essential items to be completed. It is also easy to check progress and accuracy and to use for future reference.

The order in which the chapters are tackled is flexible and can be selected according to facilities and preferences. One little tip though, it is best to complete the missing parts in pencil – just in case of mistakes!

The editor and authors wish you every success with the book and your career. Naturally if you would like to make any comments (especially constructive ones!) about the book, we would be delighted to hear from you. You can always contact the team via the publishers.

Roy Brooks
Editor

Acknowledgements

The editor, authors and publishers would like to thank all who helped so generously with information, assistance, illustrations and inspiration. In particular the book's principal illustrator, Harvey Dearden (previously principal lecturer in Motor Vehicle Subjects, Moston College of Further Education); colleagues of the Burnley College and the North Manchester College; Raymond Glancy, lecturer in Automobile Engineering at the Warrington Collegiate Institute, Warrington; and the persons, firms and organisations listed below. Should there by any omissions, they are completely unintentional.

A-C Delco Division of General Motors Ltd
Autoclenz Ltd
Autodata Ltd
Automobile Association
Automotive Products plc
Avon Rubber plc
BBA plc
Robert Bosch Ltd
Bridgestone/Firestone UK Ltd
British Standards Institution
Butterfield Equipment Ltd
Castrol (UK) Ltd
CAV, Lucas Ltd
Champion Automotive (UK) Ltd
Chrysler Ltd
Citroën UK Ltd
City & Guilds of London Institute
Clayton Dewandre Ltd
Dunlop Ltd
Fiat Auto (UK) Ltd
Ford Motor Co. Ltd
Girling Ltd
Hepworth & Grandage Ltd
HMSO
Holset Engineering Co. Ltd
Honda UK Ltd
T.F. Keller & Sons Ltd
Land Rover Ltd

Lotus Group of Companies
Lucas Varity plc
MAN Truck and Bus UK Ltd
Mercedes Benz UK Ltd
Metalistic Ltd
Mitsubishi Motors (The Colt Car Co. Ltd)
Mobelec Ltd
Mobil Oil Co. Ltd
Motor Industry Training Council
Nielson (Anteco Ltd)
Peugeot Talbot Motor Co. plc
Pirelli Tyres Ltd
Renault UK Ltd
Ripaults Ltd
Rover Group Ltd
Saab Great Britain Ltd
Shell UK Ltd
Sherrat Ltd
Sykes-Pickavant Ltd
Telehoist Ltd
Toyota (GB) plc
Unipart Group of Companies
Vauxhall Motors Ltd
Volkswagen Group United Kingdom Ltd
Weber Concessionaires Ltd
Westinghouse CVB Ltd
Zenith Carburettor Co. Ltd

Coverage of Standards

QUICK CHECK UNIT GRID

VEHICLE MAINTENANCE and REPAIR

LEVELS 1 & 2

The subject material in chapters covers **Basic Essential Knowledge** for the unit areas indicated.

UNIT NUMBERS and TITLES:

No.	Unit Title
1	Contribute to Good Housekeeping
2	Ensure Your Own Actions Reduce Risks to Health and Safety
3	Maintain Positive Working Relationships
7	Repair Tyres
8	Locate Simple Electrical/Electronic Faults
10	Remove and Replace Units and Components
11	Carry Out Routine Vehicle Maintenance
12	Diagnose Non-Complex System Faults
15	Enhance Vehicle System Features
16	Overhaul Basic Electrical Units
18	Identify and Agree Customer Vehicle Needs
19	Inspect Vehicles
20	Valet Vehicles
—	Process Payments for Purchases (METO)
—	Contribute to the Security of the Workplace (DNTO)
—	Place Goods and Materials in Storage (DNTO)
—	Join Engineering Materials Through Thermal Joining Operations (EMTA)

I = Imported Unit

VEHICLE MAINTENANCE and REPAIR

	1	2	3	7	8	10	11	12	15	16	18	19	20	— (METO)	— (Security)	— (Storage)	— (EMTA)
1. Health, Safety and Good Housekeeping	●	●	●	●	●	●	●	●	●	●	●	●	●	●	●	●	●
2. Garage Organisation & Working Relationships	●	●	●	●	●	●	●	●	●	●	●	●	●	●	●	●	●
3. Road Vehicle Systems & Layouts	●	●	●	●	●	●	●	●	●	●	●	●					
4. Vehicle Electric/Electronic Systems	●	●	●		●	●	●	●	●	●		●					
5. Vehicle Valeting	●	●	●			●		●			●	●	●	●	●	●	
6. Routine Maintenance & Inspection	●	●	●	●	●	●	●	●		●	●					●	●
7. Measurement & Dimensional Control	●	●	●			●	●	●	●	●	●						
8. Interpreting Drawings, Specifications & Data	●	●	●		●	●	●	●	●	●	●						
9. Bench Fitting	●	●	●			●	●		●	●		●				●	●
10. Joining	●	●	●			●	●	●		●		●				●	●
11. Basic Workshop and Vehicle Calculations	●		●	●	●	●	●	●	●		●	●		●			

Chapter 1

Health, Safety and Good Housekeeping

This is an introductory chapter to Health and Safety and attempts to ensure that you are aware of all the basic commonsense rules relating to personal and workshop safety. Being so aware you will be able to maintain safety protection at all times.

Important points relating to specific safety items are covered in all appropriate chapters in all the books in this series.

SAFETY ACTS AND REGULATIONS

On this page are listed the most common Acts and Regulations that affect the Safety and Welfare of employees working in garages and vehicle workshops. As an employee or student you will be expected to know, in general terms, the requirements of these various Acts/Regulations to ensure your personal safety and that of others.

Factories Act 1961 inc. (Repeals) Regulations 1976
Offices, Shops and Railway Premises Act 1963
Fire Precautions Act 1971
Health and Safety at Work Act 1974

The above **FOUR** items are Acts of Parliament and are enforceable by law.

Below is a list of various Regulations which generally apply to special situations. The Regulations expand or place into detail the intentions of the Acts. They are made as the need arises and carry as much legal authority as the Act under which they are made. New or redesigned Regulations therefore keep the Acts up to date.

Asbestos Regulations 1969
Abrasive Wheels Regulations 1970
Highly Flammable Liquids and LPG Regulations 1972
Protection of Eyes Regulations 1974
Notification of Accidents and Dangerous Occurrences Regulations 1980
Safety Signs Regulations 1980
First Aid at Work Regulations 1981
Control of Substances Hazardous to Health (COSHH) Regulations 1988
Noise at Work Regulations 1989
Local Bylaws

Name **FIVE** Safety Regulations created in 1992 which implement **European Community Directives** to improve the quality of safety that may be considered relevant to the needs of the Motor Trade:

1. ....
2. ....
3. ....
4. ....
5. ....

Health and Safety at Work Act

The Health and Safety at Work Act 1974 provides one comprehensive system of law, dealing with the health and safety of working people and the public as affected by working activities. *Note:* This act has now been updated by the Management of Health and Safety at Work Regulations 1992.

State in general terms what implications are required by the individual and the company under current health and safety legislation:

....

....

....

General duties of employers to their employees

Give examples of, or expand on the list given below, the employer's responsibilities towards the maintenance of safety in terms of providing:

1. a safe place of work; e.g. ....
2. safe plant and equipment; e.g. ....
....
3. a safe system of work ....
....
4. a safe working environment
   (i) maintenance of reasonable working ....
....
   (ii) provision of adequate ....
....
5. safe methods of handling, storing and transporting goods
   (i) ....
....
   (ii) ....
....
6. a procedure for the reporting of accidents ....
7. information ....
8. a safety policy ....
....

General duties of employees

State THREE duties of employees with regard to health and safety at work:

1. ....
   ....
   ....

2. ....
   ....

3. ....
   ....

Give an example of a situation where an employee could be prosecuted under the terms of the Act:

....

....

....

Enforcement

A body called The Health and Safety Executive Inspectorate enforce the Act. Its inspectors have various powers and penalties at their disposal.

What procedures can be adopted by the inspectorate?

....

....

....

....

....

....

Also Local Authorities appoint Environmental Health Officers who inspect Offices and Shops.

BASIC PRINCIPLES OF ACCIDENT PREVENTION

An accident is an unexpected, unplanned occurrence which can involve injury. The effects of an accident are not necessarily confined to the person or persons directly involved; other people may be affected in different ways. A minor accident may have trivial effects but a serious accident could affect one's whole life socially, domestically and economically. Accident prevention should therefore be a major concern at all times.

Indicate In the table below, how an accident, where, for example, a man loses a limb, could affect the persons listed.

People affected	Possible effects
Person injured	....
Service manager	....
Foreman	....
Workmates	....
Parents or family	....

Describe ways in which a positive personal attitude should be developed in order to prevent accidents:

1. ....
   ....
   ....

2. ....
   ....
   ....
   ....

Causes of Accidents

Generally speaking accidents are caused by:

Human

1. Ignorance of the dangers involved.
2. Failure to take adequate precautions.
3. Tiredness causing lack of concentration.
4. Fooling about.

Environmental

1. Unguarded or faulty machinery.
2. Incorrect or faulty tools.
3. Inadequate ventilation.
4. Badly-lit workshops.

State **SIX** other causes of accidents in the workplace:

1. ....
2. ....
3. ....
4. ....
5. ....
6. ....

Accident prevention

What are the requirements for successful accident prevention?

1. ....
2. ....
3. ....
4. ....
5. ....
   ....
6. ....
   ....
7. ....
   ....

Near Miss Accident Prevention

In considering the definition of 'Accident Prevention' you must look at what might have happened just as, or more seriously than, what actually did happen. A 'Near Miss' is what is likely to happen the next time you are in the same situation as when the 'Near Miss' occurred.

The drawings below may be amusing but the near misses are potentially more serious accidents than the ones that occurred.

Describe any accident or 'near miss' accident that you have experienced and state how it could have been avoided:

....
....
....

SAFETY SIGNS

See current UK and European Safety Legislation. A safety sign gives a particular message about health and safety by a combination of its (1) geometric form, (2) contrasting colours and (3) a symbol or text.

The Geometric form indicates:

A circular band and cross bar ....

A triangle with a thick border ....

A circle ....

A rectangle ....

A Safety colour indicates:

Red ....

Yellow ....

Blue ....

Green ....

Contrasting colours make the sign more conspicuous.
Black is used with Yellow. White is used with Red, Blue or Green Safety colours.

PROHIBITION SIGNS: A Black symbol on a White background with a Red circular band and cross bar.

State what prohibition is indicated:

CAUTION SIGNS

A Yellow background with a Black symbol and triangular border.

State the risk that each condition indicates:

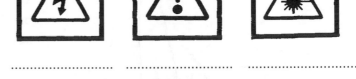

MANDATORY SIGNS

A White profile on a Blue circular background.

State the type of protection that must be worn:

5

PERSONAL PROTECTION

See Personal Protection Equipment at Work (PPE) Regulations 1992.
Many accidents are caused by the negligence of the person or persons involved. One very important factor with regard to accident prevention is the degree of 'personal protection' in relation to the particular hazards involved.

Personal protection can range from the use of adequate working clothes to the use of specialised protective equipment. All Personal Protective Equipment (PPE) in use at work should carry the CE mark and where appropriate should comply with a European Norm (EN) standard. The following headings all relate to personal protection. Describe the protective measures to be adopted, under each of the headings, for garage workers.

Ordinary working clothes (that is, protective clothing)

Two essential items of protective clothing are overalls and safety footwear.

....

....

....

....

....

Eye protection

Special types of goggles must be worn which are to CE and BS Specifications when carrying out certain work operations. Examples:

....

....

....

....

....

Hearing protection

Hearing protection must be used if working on an area constantly having a decibel reading of 90+ dB(A)

....

State the personal protection safety features indicated on the drawing below:

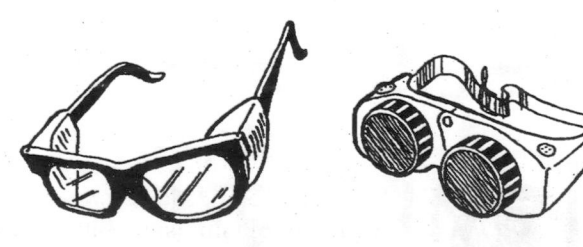

Personal presentation

This includes personal hygiene, use of personal protection equipment, clothing and accessories suitable for the workplace.

Your personal presentation at work should:
- help to ensure the health and safety of yourself and others
- meet legal requirements
- be in accordance with workplace policies.

What would the types of eye protection shown below be used for?

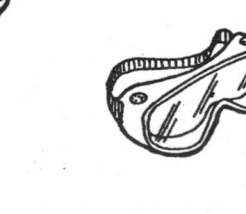

....

6

Eye Protection

See Protection of Eye Regulations 1974.

Complete the table below by stating the type of eye protection required.

GENERAL HAZARD	SPECIFIC HAZARD	EYE PROTECTION
MECHANICAL HAZARDS e.g. sparks, dust or flying particles	Machining	....
	Grinding, chipping, cutting	....
CHEMICAL HAZARDS e.g. splashes, fumes or burns	Handling chemicals	....
THERMAL and RADIATION HAZARDS	Gas welding	....
	Spot welding	....
	Electric arc welding	....

You only have one pair of EYES

Ear Protection

Under the Noise at Work Act 1989, both employers and employees have obligations which they must fulfil to minimise the risk of damage to hearing in the workplace. Below is a summary of the employer's duties; state the employee's duties and tick the items where each must comply.

EMPLOYER'S DUTIES	Below 85 dB(A)	At first Action Level – 85 dB(A)	At second Action Level 90 dB(A)
1. To reduce the risk of hearing damage to the lowest level reasonably practicable			
2. To appoint a competent person to carry out noise assessments and keep a record of those assessments			
3. To reduce the risk of exposure to noise as far as is reasonably practicable by means other than ear protectors			
4. To provide adequate information, instruction and training about risks to hearing, and how to minimise those risks. To provide information on how to obtain ear protectors for levels at 85 dB(A) and at 90 dB(A), and employer's obligations under the regulations			
5. To mark ear protection zones with notices as far as is reasonably practicable			
6. To ensure that ear protectors are:			
(a) provided to employees who ask for them			
(b) provided and used by all exposed			
(c) maintained and repaired			
7. To ensure that all who go into ear protection zones are wearing ear protectors			
EMPLOYEE'S DUTIES			
Employees must so far as practicable:			
(a)			
(b)			

You only have one pair of EARS

Head protection

Three types of head protection may be used in a garage workshop. These are:

1. cotton or wool caps

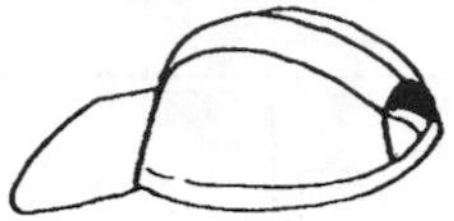

....
....

2. hair net cap

....
....
....

3. bump cap

....
....
....

Skin protection

Thoroughly cleaning the skin, particularly hands, face and neck, is extremely important. How should hands be protected?

....
....
....
....
....

Hand protection

Give TWO examples of when gloves should be worn. 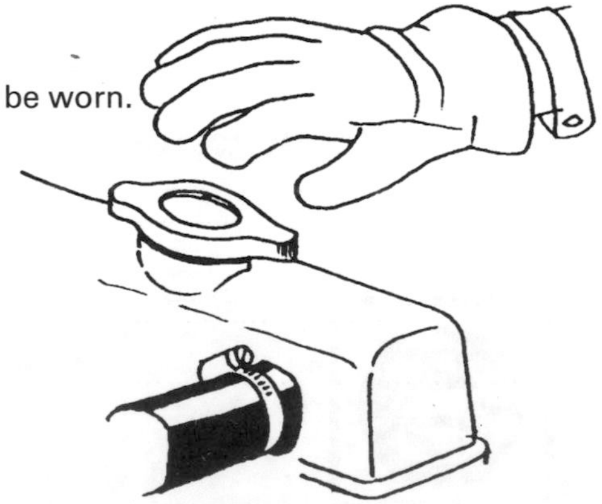
1. Heat resistant gloves should be worn:

....
....
....

2. Plastic gloves should be worn:

....
....
....

SAFE USE OF MACHINERY AND EQUIPMENT

Many accidents in garages are caused either by the employee not taking adequate precautions or by faulty equipment.

If you were asked to work on the vehicle shown below, what TWO precautions would you insist upon before starting?

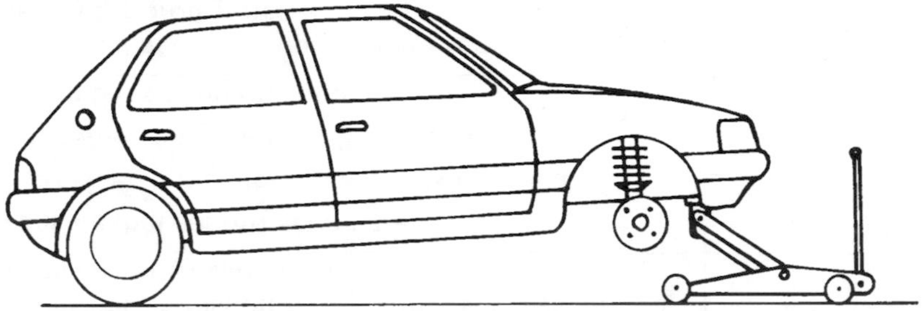

1. ....
2. ....

State TWO other precautions that should be observed:

1. ....
2. ....

State precautions, other than those shown, that are necessary when working on a vehicle raised by a ramp (hoist):

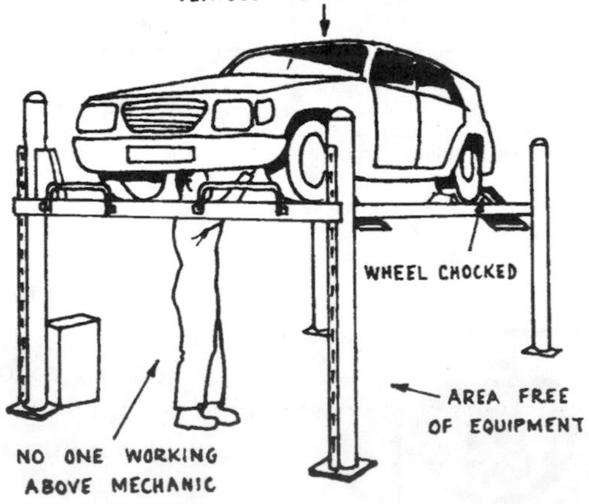

....
....
....
....
....
....
....
....
....

VEHICLE CENTRAL ON HOIST

WHEEL CHOCKED

AREA FREE OF EQUIPMENT

NO ONE WORKING ABOVE MECHANIC

What precautions should be taken when using the following items of equipment?

Compressed-air equipment

....
....
....
....
....

Chain lifting blocks

....
....
....
....

Which is the correct way to lift an engine using a chain sling as shown below?

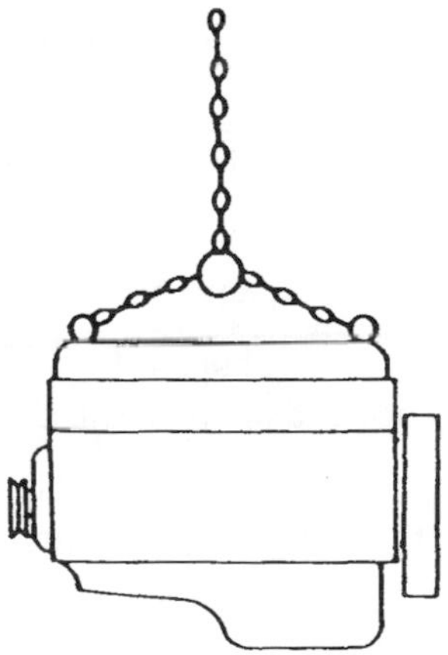

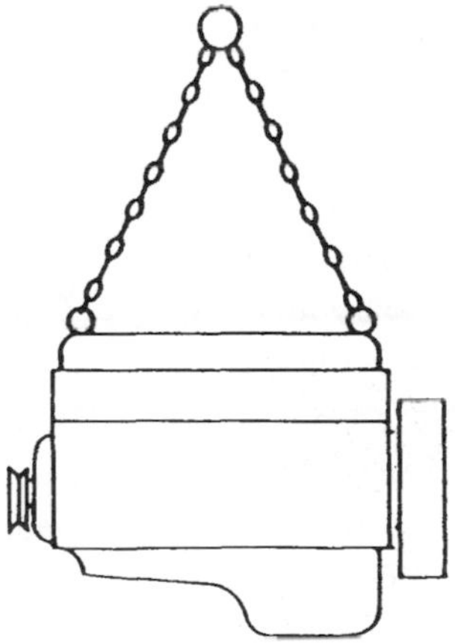

....

Slings and chains should be checked for wear at least once every ....

Grinding wheels

....
....
....

Bench drills

....
....
....
....

Tidiness

....
....
....
....
....
....
....

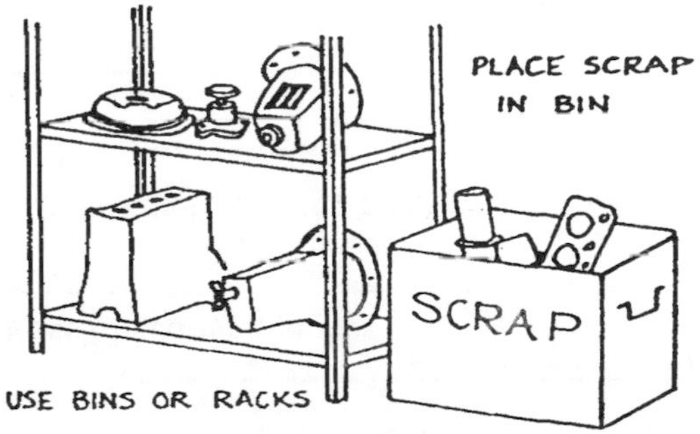

PLACE SCRAP IN BIN

SCRAP

USE BINS OR RACKS

HAZARDS AND RISKS

The Health and Safety Executive (HSE) is the body appointed to support and enforce health and safety law. They have defined **two** important concepts as follows:

Hazard 'a hazard is something with potential to cause harm'.

Risk 'a risk is the likelihood of the hazard's potential being realised'.

Almost anything may be a hazard, but may or may not become a risk.

Give examples below of major hazard or risk areas associated with a garage/workshop environment.

....

....

....

....

....

....

....

....

....

....

....

....

....

....

Indicate below the possible injuries that could result from the situations listed.

Situation		Possible consequences
Poor spanner fit on nut		....
Undue effort needed owing to the use of short spanner		....
Using a file without a handle		....
Using a blunt screwdriver		....
Using a 'mushroomed' headed chisel		....
Banging two hammer faces together		....
Using a file as punch		....

COSHH Regulations 1988

Under COSHH or the Control of Substances Hazardous to Health Regulations 1988, all persons at work need to know the safety precautions to take so as not to endanger themselves or others through exposure to substances hazardous to health. Below are four general classifications of risk – know the appropriate symbols, and their safety precautions. State the meaning of each symbol.

MEANING		SAFETY PRECAUTIONS
TOXIC/VERY TOXIC	☠	1. Wear suitable protective clothing, gloves and eye/face protection. 2. After contact with skin, wash immediately with plenty of water. 3. In case of contact with eyes, rinse immediately with plenty of water and seek medical advice. 4. In case of accident or if you feel unwell, seek medical advice immediately.
CORROSIVE	(corrosive symbol)	1. Wear suitable gloves and eye/face protection. 2. Take off immediately all contaminated clothing. 3. In case of contact with skin, wash immediately with plenty of water. 4. In case of contact with eyes, rinse immediately (for 15 minutes) with plenty of water and seek medical advice.
HARMFUL	✖	1. Do not breathe vapour/spray/dust. 2. Avoid contact with skin. 3. Wash thoroughly before you eat, drink or smoke. 4. In case of contact with eyes, rinse immediately with plenty of water and seek medical advice.
IRRITANT	✖	1. In case of contact with eyes, rinse immediately with plenty of water and seek medical advice. 2. In case of contact with skin, wash immediately with plenty of water. 3. Do not breath vapour/spray/dust. 4. Avoid contact with skin.

COSHH Regulations 1988 have placed legal obligations on employees in the work environment. The main aim of the Regulations is to ensure that the exposure of an employee to substances hazardous to health is prevented or where this is not reasonably practical, adequately controlled.

State as an employee THREE obligations required by you to ensure maintenance of good health.

1. ....
   ....
   ....

2. ....
   ....

3. ....
   ....

In addition it should be incumbent on you to:
4. present yourself during working hours for medical surveillance if requested to do so by your employer.
5. supply information about your health during medical surveillance as the medical adviser or appointed doctor may reasonably require.
6. remove contaminated personal protective equipment and store in an appropriate place before eating or drinking.

PAINT

Many paints used in a garage contain chemicals which present health hazards if sensible precautions are not taken.
NOTE: Long-term effects do not immediately affect your health.

State opposite each symbol what sensible precaution should be taken when using paint:

11

Harmful Substances

Certain areas in a motor-vehicle repair premises present particular health hazards. The hazards may, for example, be due to breathing in polluted air or coming into contact with harmful substances. (See COSHH Regulations.)

What is meant by the term *toxic* when used to described a substance or gas?

....

....

....

....

....

List some of the toxic gases or substances likely to be present in a motor-vehicle repair workshop:

....

....

....

....

....

....

....

Asbestos dust may be encountered when cleaning brake and clutch assemblies. What hazard is associated with asbestos dust and what precautions must be taken?

....

....

....

....

Complete the table below by describing the toxic hazards involved in the areas listed and briefly outline the precautions to be adopted.

Hazard area	Hazard	Suitable precautions
Engine tuning	*Exhaust fumes*	*Pipe gases outside, adequate* .... *ventilation, use of extractor fans, gas* *not aimed into confined space*
Welding bench	....	....
Degreasing plant	....	....
Body shop	....	....
Paint shop	....	....
Battery charging	....	....

Under what motor-vehicle workshop conditions would it be considered necessary to wear a fresh air fed respirator?

....

....

Electrical Safety

Two dangers arising as a result of using electricity in a workshop are: *fire* caused by say overheating of an electrical circuit or a bursting bulb igniting fuel, and *electric shock* as a result of someone coming into contact with a live circuit.

For safety reasons hand-held electrically operated equipment, for example portable drills, should be .... volts and handlamps should be .... volts.

What could cause a circuit to overheat?

....

....

....

....

....

....

....

....

....

The main hazards arising from the use of electrical equipment are:

(a) *Poor or damaged insulation on such as cables, plugs, etc.* ....

(b) *Lack of adequate earthing for the equipment.* ....

(c) ....

(d) ....

(e) ....

(f) ....

An electric shock can cause death or serious injury to a person. Apart from electric shock, what other effects can result from a person coming into contact with a live electrical conductor?

....

Electric shock is caused by current passing through the human body; how does this occur?

....

....

....

....

....

....

Describe briefly the action to be taken in the event of a person receiving an electric shock:

....

....

....

....

....

....

....

....

....

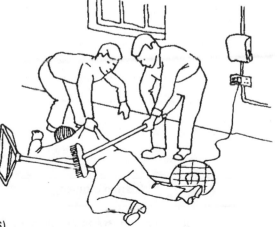

Note: See First-Aid section (page 16).

Flammable Liquids and Gases

Many flammable substances are used in garages, for example

(a) _Petrol_ .... (c) .... (e) ....

(b) .... (d) .... (f) ....

Some liquids are _volatile_. What is meant by this and what particular hazards can this present during the normal course of repair work?

....

....

....

....

....

....

....

What procedure should be adopted in the event of a spillage of flammable liquid such as petrol?

....

....

....

....

....

....

....

Some accidents related to flammable liquids or gases are shown in the table below. Complete the table by stating possible causes of the accident.

Accident	Possible cause
Petrol tank explosion	....
Battery explosion	....
Fire in the pit	....
Fire under a bonnet	....
Fire at or near welding bench	....

What special precautions should be taken with regard to the storage of

flammable substances? ....

....

....

....

....

....

What substance is the more dangerous with regard to storage, a substance with a high flash point or a substance with a low flash point?

....

FIRE FIGHTING EQUIPMENT

Liquids or chemicals that are highly combustible are commonly found in motor-vehicle workshops, for example, petrol, cleaning solvents, paints, etc. When not being used, they must be kept in fire-proof containers and these must prevent leakage and evaporation. The vapour from these chemicals can be ignited by a spark as small as from a light switch.

It is important therefore that everyone not only tries to prevent fire, but also has a working knowledge of the use and correct type of fire extinguisher required to eliminate a fire.

Complete the triangle to show the three conditions that must be present simultaneously in order to start a fire and one action needed to stop it.

State the **TYPES OF FIRE** under the Classification of Fire Risk, and **TICK** in the grid which type of FIRE Extinguisher would be suitable.

CLASSIFICATION OF FIRE RISK	WATER	FOAM	CO₂ GAS	POWDER	BCF
A					
B					
C					
⚡					
🚗					

All fire extinguishers must be red and contain a colour code label in accordance with the European Standard BS EN3: 1996.
State the colour code label for the fire extinguishers shown above.

Foam ....

Water ....

CO_2 gas (carbon dioxide) ....

Vaporising liquid BCF ....

Dry chemical powder ....

The main chemicals used in fire extinguishers are listed below. Explain why they are suitable for fighting fires.

Water ....
....
....
....

Foam ....
....
....

CO_2 gas (carbon dioxide) ....
....
....

Dry powder ....
....
....
....
....
....
....

Inert (vaporising liquid) BCF
....
....

Fire Prevention and Control

Protection against fire is normally organised in accordance with the requirements of the Factory Acts and in co-operation with the local Fire Prevention Officer. Fire-fighting equipment must be readily available and kept properly maintained. Doors and passages must be kept clear and a positive routine established, to be followed in the event of a fire.

Briefly describe the procedure to be followed in the event of a fire in the workshop:

....

....

....

....

....

....

....

....

Investigation

What are the visible signs of fire prevention in your workshop?

....

....

....

Water from a hose, bucket or extinguisher is used on solid fuel fires. What effect has the use of water on burning flammable liquids?

....

....

List the types of extinguisher available in the college workshop.

....

....

FIRST AID

Personnel should be familiar with the location and contents of First-Aid boxes. Cuts, abrasions and burns, etc., however minor, should be cleaned and treated promptly owing to the dirty nature of motor-vehicle repair work.

Certain personnel trained in basic First Aid should be available to provide treatment and advice during working hours; the staff should know how and where to contact these people promptly in the event of an accident.

Basic First Aid

By reference to the British Red Cross Society First Aid Chart, briefly describe the procedure to be followed when dealing with the following accidents:

Bleeding

....

....

....

....

....

....

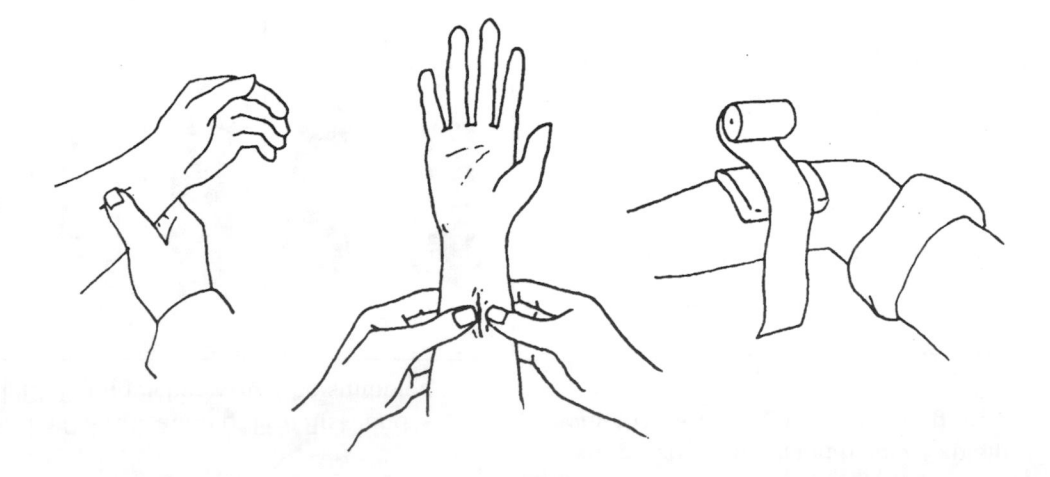

Fractures

....
....
....
....
....
....
....

Burns and scalds

....
....
....
....
....

Hand burnt   **Chemical burns**

Unconsciousness

....
....
....
....

Breathing stopped

....
....
....
....
....
....
....
....
....
....

....
....

Accident reporting

Accidents of any kind should be reported to the employer. Describe normal procedure for reporting and recording accidents.

....
....
....
....

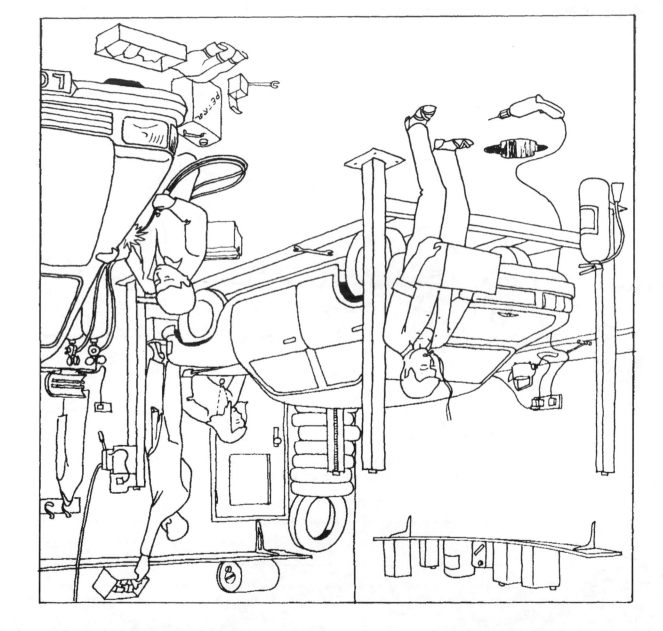

GARAGE WORKSHOP HAZARDS

Examine the drawing and list the hazards you can find in the garage workshop shown:

WORKSHOP INSPECTION

To maintain good housekeeping, make a thorough inspection of your workshop or garage and make brief notes under the headings on this page to describe any potential safety hazards or lack of warning or guidance notices:

Fire risk/fire precautions

....

....

....

Machinery (drills, grindstones, etc.)

....

....

....

Vehicle lifts and jacks

....

....

....

Electrical (hand drills, hand lamps, etc.)

....

....

....

Lifting equipment

....

....

....

Welding area

....

....

....

Battery charging

....

....

....

Compressed-air equipment

....

....

....

General tidiness (floor condition, etc.)

....

....

....

State where in the Garage/Workshop the following can be found:

First-Aid equipment ....

....

a designated First Aider (Name) ....

....

location of Fire Extinguishers and types ....

....

position of Machine Isolaters ....

....

MOVEMENT OF LOADS

Any heavy object which requires moving manually or by mechanical lifting equipment is considered to be a load.

In a large garage or parts department, heavy loads may be transported in the manners shown. Name each method of transport.

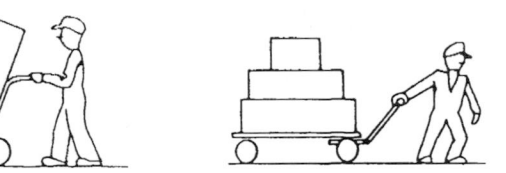

1. ....   2. ....   3. ....

Which of the above units is loaded correctly? ....

Below is shown a wheeled stand. When would such a unit be used for moving loads in a workshop?

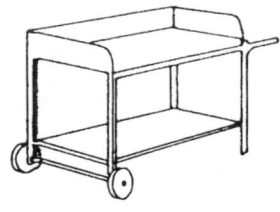

....
....
....
....
....

Show with the aid of simple sketches, methods of chocking to avoid overbalance or shifting.

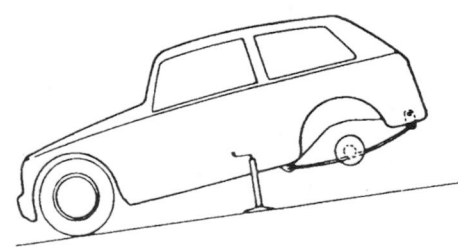

State four typical items that may need to be physically carried from the garage workbench to the vehicle:

1. ....   3. ....

2. ....   4. ....

Manual Handling of Loads

More than a quarter of accidents reported each year are associated with manual handling – the transporting or supporting of loads by hand or by a bodily force – and the most common type of strain or sprain is a back injury.
When considering a LOAD that requires moving, how could the possibility of reducing the risk of injury be improved?

1. ....

2. ....

3. ....
....

4. ....

5. ....
....

How should the working environment be modified to reduce the risk of injury?

1. ....
....
....

2. ....
....

3. ....

4. ....

5. ....

If an item or items are to be moved manually, what precautions and inspections should be made before actually lifting?

1. ....

2. ....

3. ....
....

4. ....

5. ....

Lifting

ONE PERSON LIFT (SQUAT LIFT)

State the lifting methods indicated:

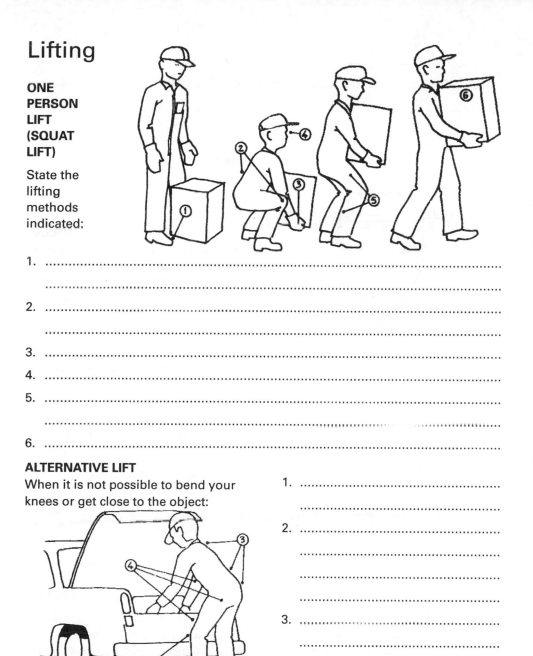

1. ....
....
2. ....
....
3. ....
4. ....
5. ....
....
6. ....

ALTERNATIVE LIFT

When it is not possible to bend your knees or get close to the object:

1. ....
....
2. ....
....
....
....
....
3. ....
....
4. ....

UNLOADING

1. ....
....
2. ....
....
3. ....
....
....
4. ....
....

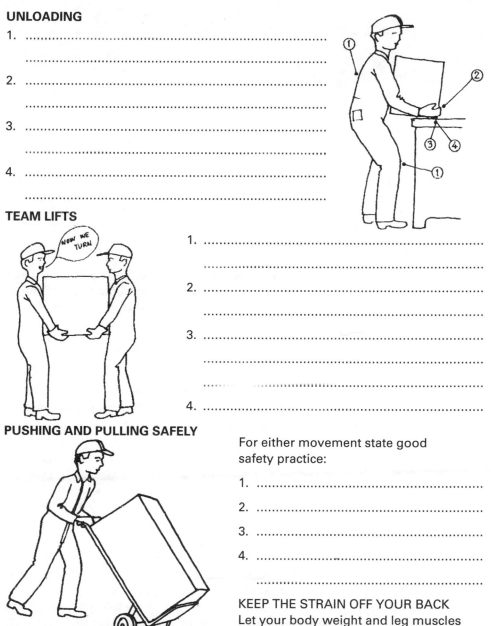

TEAM LIFTS

1. ....
....
2. ....
....
3. ....
....
....
4. ....

PUSHING AND PULLING SAFELY

For either movement state good safety practice:

1. ....
2. ....
3. ....
4. ....
....

KEEP THE STRAIN OFF YOUR BACK
Let your body weight and leg muscles do the work.

Lifting Gear Accessories

When removing an engine from a vehicle the accessories required to support the engine to the lifting frame would be hooks, chain or sling, and eye bolts or shackles.

Examine your lifting equipment at work and note what is provided.

Lifting hook

Typical
safety
hook

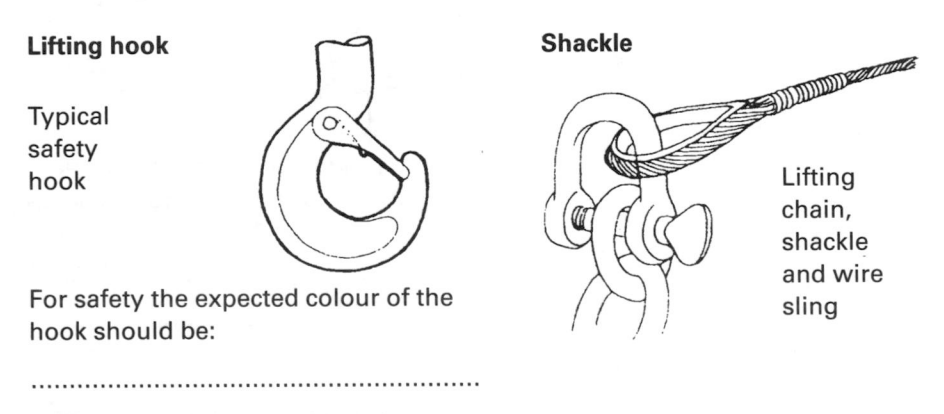

For safety the expected colour of the hook should be:

....

Shackle

Lifting
chain,
shackle
and wire
sling

Position of chains and slings

The angle made by the slings is very important. What is the maximum recommended angle (A) between the slings? ....
If the angle (A) were to be substantially increased, what would be the effect of the 'pull' on the slings?

....

....

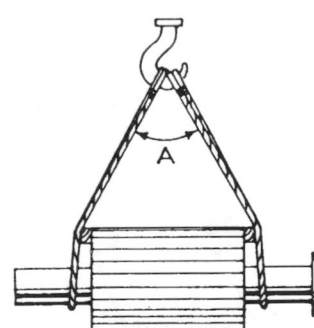

State FOUR rules to be observed when using slings:

....

....

....

....

Lifting devices

Lifting devices may be classified by their power sources. State the power source of the lifting devices shown:

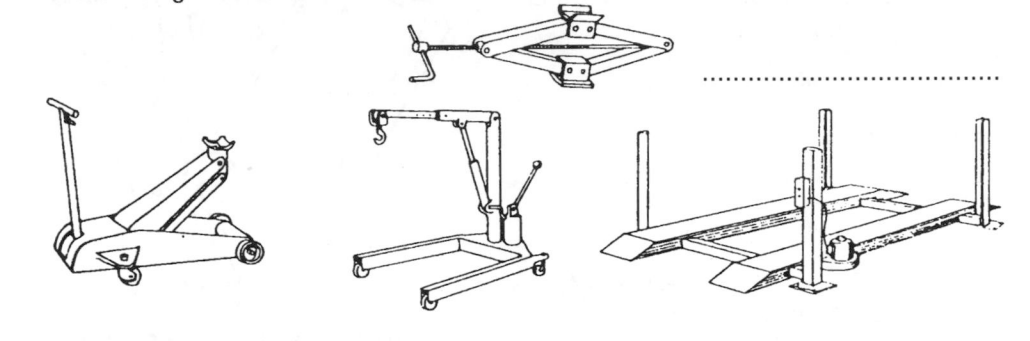

....

....

Above is shown basic garage lifting equipment. What other power lifting sources are available even though they are not commonly found in a garage workshop?

....

....

As a general rule any load over .... requires some form of powered lifting gear to support or move it. State the safety rules that should be observed when moving a swinging load across the workshop:

....

....

....

....

....

....

....

....

....

MAINTAIN THE CLEANLINESS OF MACHINERY, EQUIPMENT AND WORK AREAS

Why is it necessary to keep workshop machinery, equipment and the work area clean?

....

....

All workshop equipment should be cleaned before carrying out routine maintenance operations.

State examples when items generally require cleaning.

MACHINERY

1. Stand drill ....
2. Hydraulic press ....
3. Compressor ....

SPECIALISED EQUIPMENT

1. Vehicle ramps (Hoist) ....
2. Engine jib crane ....
3. Oil drain containers ....
4. Hand tools ....

WORKSHOP AREAS

1. ....
2. ....
3. ....

State what safety precautions must be observed before cleaning machinery, equipment and the workplace:

....

....

....

....

....

....

State the types of cleaning materials that are suitable and authorised for cleaning machinery, equipment and the work area:

PERSONAL

Barrier cream ....

Hand cleaning material ....
Some modern garage workshops include showers as part of their personal cleaning facilities.

MACHINERY

Stand drills ....

Valve grinders..

Cleaning tanks ....

EQUIPMENT

Vehicle ramps ....

Trolley jacks..

Special and personal tools ....

WORKSHOP

Complete floor area ....

Individual car bays..

Walls ....

Lighting ....

TYPICAL CLEANING/SAFETY EQUIPMENT

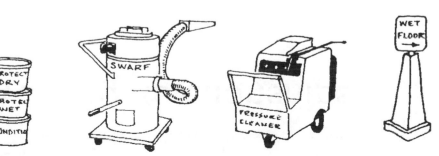

23

Chapter 2

Garage Organisation and Working Relationships

THE MOTOR TRADE

Goods being distributed from a manufacturer or producer to the consumer usually flow through a distribution network which is as follows:

MANUFACTURER → .... → .... → CONSUMER

In the motor industry some vehicles, parts and accessories reach the consumer via the following channels.

VEHICLE/PARTS MANUFACTURER (PRODUCER)
↓
DISTRIBUTOR   (.......................................)
↓
DEALER   (.......................................)
↓
CONSUMER   (.......................................)

Name two distributors and two of their respective dealers in your area; state the manufacturer in each case.

1. Manufacturer ....   2. ....

   Distributor ....

   Dealer ....

In the motor trade there has been a move to delete one tier of the sales distribution network. Thus many car sales are likely to be

Manufacturer → Dealer → Customer

State two likely consequences of this move:

....

....

....

....

Name two manufacturers and two of their respective direct dealers in your area:

Manufacturer   1. ....   2. ....

Dealer   1. ....   2. ....

List the main services provided by the sales and servicing side of the motor industry:

New and used vehicle sales ....

....

....

....

....

....

....

....

....

The main departments in a typical dealership are:

New and used vehicle sales ....

....

....

....

....

....

STAFF ROLES

A garage workshop may employ TECHNICIANS, CRAFTSMEN and OPERATIVES.
Describe briefly the role of each type of worker.

TECHNICIAN

....

....

....

....

....

CRAFTSMAN

....

....

....

....

....

OPERATIVE

An operator would be involved in relatively simple, very often repetitive activity,

e.g. inspection, machining, packing/unpacking, cleaning, fork lift truck driving, etc.

....

....

WORKSHOP FOREMAN

His role is to allocate the jobs, liaise with the service manager and reception,

supervise and inspect work in progress, ensure safe working practices are

adopted and generally ensure that management policies are implemented.

State briefly the roles of the garage personnel listed below:

SERVICE MANAGER ....

....

....

....

....

....

RECEPTIONIST ....

....

....

....

....

CAR SALES STAFF ....

....

....

PARTS DEPARTMENT STAFF ....

....

....

ADMINISTRATIVE STAFF ....

....

....

ORGANISATIONAL STRUCTURE

The levels of authority, duties and responsibilities, and working relationships of all personnel and departments within a company should be clearly identified and understood by employees at all levels.

A structure diagram such as the one shown below can indicate levels of authority and responsibilities of staff.

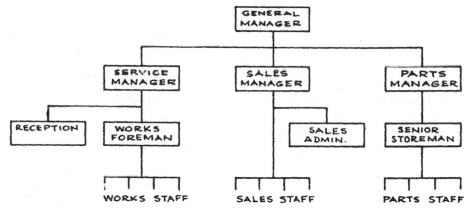

What do the vertical and horizontal links indicate on the diagram?

....
....
....
....

Describe briefly what is meant by:

1.   JOB SPECIFICATION

....
....

2.   CONTRACT OF EMPLOYMENT

....
....
....
....

Write out your own job description.

JOB TITLE

....

DUTIES

....
....
....
....

RESPONSIBLE TO:

....

COMMUNICATION

In many instances, problems and misunderstandings arise as a result of inadequate communication.

An essential factor contributing to the smooth efficient operation of an organisation is the establishment of clearly defined *lines of communication*.

State the methods of communication employed within an organisation when dealing with customers and colleagues:

....

Give examples of the use of such methods of communication:

....
....
....
....
....
....
....
....

WORKING RELATIONSHIPS AND TEAMWORK

Positive working relationships develop with the people you interact with at work. List some people with whom you will need to maintain a positive working relationship.

....

....

....

....

....

....

Good working relationships are very important to the success of every business.

Working as a Team

To make the company successful, all of its employees must work together. They must **co-operate**, like members of a football team: this is **teamwork**.

If the firm does well, and employees get on with each other and trust each other, there will be a good feeling in the workplace, and people will be enthusiastic about their jobs. This feeling is called **good morale**.

When everyone works hard, and no one wastes time or resources, the firm will be **efficient**. By being efficient, employees will get a lot done – they will be **productive**.

If a team with good morale works efficiently and productively, customers will be satisfied and pleased to come again. They are also likely to recommend the firm to others, so it will gain a good **reputation**. And this in turn will bring more business, and the firm will become even more successful, and will grow. It will gain a good **company image**.

Building Good Relationships

Some things which contribute to good working relationships are:

(a) working well as a team, cooperating and sharing ideas;

(b) using the correct method of communication which is appropriate to the particular situation;

(c) being honest and fair;

(d) recognising that there are differences in personality and temperament in colleagues;

(e) understanding your own and your colleagues job role;

(f) ensuring that work targets are sensibly achievable;

(g) adopting safe working practices and adhering to rules and regulations.

List some of the things which could upset good working relationships.

....

....

....

....

....

....

....

....

....

CUSTOMER RELATIONS

Customers provide the business and profit necessary for the survival of a company. Bearing this in mind, staff at all levels from workshop labourer to senior executive should be made well aware of the IMPORTANCE OF THE CUSTOMER.

List the factors which have a direct bearing on good customer relations:

Customer image
....

Overall appearance of premises
....

Administration and organisation of company
....

....

....

....

....

The experience a customer goes through when dealing with a firm should ensure that he or she 'stays sold', that is, he or she will continue to be a customer. The establishment of a long lasting relationship with customers or customer retention (goodwill) is a basic foundation of a successful business enterprise.

The company image can be enhanced by promoting the resources available, for example:
State of the art equipment (diagnostic – body repair – paint facility, etc.)
Extensive parts department, vehicle sales showroom
Staff expertise

When a prospective customer makes his or her first contact with a company, either in person or by telephone, the first impression should always be favourable; there should be no cause of irritation.

List some of the things which could be irritating to a prospective customer when making initial contact by telephone:

Lengthy ringing time before response
....

....

....

....

Information regarding day-to-day queries should always be at hand for the customer, or details can be taken and a promise made to communicate information by telephone or post as soon as possible.

In what ways can the business premises give a favourable impression to a prospective customer?

Ease of access and parking
....

....

....

....

....

RECEPTION – CUSTOMER VEHICLE NEEDS

Customers requiring vehicle service or repair will be dealt with in the reception area.
State the functions of customer reception:

To greet the customer
....

To provide a comfortable, pleasant and friendly environment for the customer
....

To listen and take note of customer requirements
....

To advise the customer
....

....

....

....

....

....

The reception staff should therefore have knowledge of:

Customers: business, private, informed, non-informed, new, regular
Customer requirements
Vehicle types and specifications
Service and repair operations
Diagnostic techniques
Costs
Timescales
Availability of labour
Availability of necessary staff skills
Availability of tools, test equipment and service facilities
Availability of replacement parts
Other work priorities

Manufacturers' warranties that may be applicable

Give some examples of the things which may influence a customer one way or the other when dealing with employees:

(a) *The visual appearance of an employee.*

(b) *The approach to customers, e.g. correct greeting, polite, pleasant, etc.*

(c) *Good clear speech.*

The way in which customer complaints are handled can either boost or be damaging to customer relations. The approach or technique adopted when dealing with a complaining customer will vary according to the type and attitude or behaviour of the customer. For example, complaining customers may be angry, confused, demanding, embarrassed, pleasant, etc.

List some of the causes of customer complaints:

Faulty goods

Price

Quality of work or service

Complaints may be justified or unjustified. However a customer may not be aware of the fact that his or her complaint falls into the last mentioned category.

List some of the important points to be aware of when dealing with a complaint:

(a) *Adopt a positive attitude.*

(b) *Listen carefully, obtain the facts correctly, and try to establish the validity of the complaint.*

FAIR TRADING

Under the terms of the *Fair Trading Act*, it is illegal to adopt trade practices which adversely affect the economic interest of the consumers. To enforce this, the local Trading Standards Officer can enter motor-trade premises, make test purchases and inspect goods etc.

To what does 'Trade Practice' relate?

(a) *Advertising, labelling and packaging of goods.*

There is often a huge difference between what the customer expects and the service he or she receives. Narrowing the gap is essential to staying in business. Closing the gap, that is meeting or even exceeding the customers' expectations, should be the aim if customer satisfaction and loyalty are to be gained.

List three things which would have a detrimental effect on customer relations:

....

....

....

In an attempt to monitor performance relating to customer standards most manufacturers conduct surveys throughout their dealership network and produce a points scheme for the dealerships. Incentive rewards are often given to dealerships achieving top results.

Shown on this page is an example of a small section of a Customer Standards Survey, as operated by Renault.

CUSTOMER STANDARDS SURVEY

Reception

1-1	How many rings did it take to answer the telephone?	Opposite is shown a
1-2	Did the Receptionist say 'Good Morning/Afternoon'?	particular dealership
1-3	Did the Receptionist give the name of the Dealership?.........	score and ranking
1-4	Did the Receptionist offer you assistance?...........................	in an area of 24
1-5	Did the Receptionist ask for your name?	dealerships

Transfer

1-6	How many seconds did it take to transfer you?....................	The part of the survey
1-7	Did they use your name on answering?..............................	shown relates to
1-8	Did they say the name of the department?	After Sales Service
1-9	Did they tell you their name?	
1-10	Did they offer to book your car in?	
1-11	Did they use your name at any stage?................................	

Call Back

1-12 Were you called back within 30 minutes of the agreed time? ....

1-13 Did they know what your original enquiry was about?.........

1-14 Did they tell you their name? ....

(1) Being made to feel welcome
(2) Getting an appointment within a reasonable period
(3) Prompt attention when you arrived with your vehicle
(4) Advice about the work needing to be done
(5) The attention paid to any minor details you had reported
(6) The work being completed when promised
(7) The completion of all the jobs you requested
(8) The quality of work carried out
(9) The cleanliness of the vehicle when returned
(10) The availability of reception staff when you needed them
(11) The helpfulness and politeness of dealership personnel
(12) The explanation of the invoice
(13) The invoice corresponding with the work you had requested
(14) The cost represented value for money

	Your points	Area average points	Ranking
(1)	95.45	89.93	7/24
(2)	100.00	90.74	1/24
(3)	95.45	88.51	5/24
(4)	85.37	75.97	5/24
(5)	82.86	65.53	4/24
(6)	86.36	80.44	8/24
(7)	90.00	66.53	4/24
(8)	90.91	74.06	5/24
(9)	95.45	84.89	4/24
(10)	100.00	89.52	1/24
(11)	95.45	92.35	10/24
(12)	84.62	82.35	11/24
(13)	90.00	86.40	6/24
(14)	20.00	37.08	20/24
Overall	86.57	79.02	4/24

COSTING AND CHARGING

Before a customer can be charged for a particular job all the costs incurred in doing the job must be determined.

A STANDARD charge can be applied to certain jobs, for example some manufacturers will cost and issue publications showing standard charges for specific jobs.

Give examples of jobs which could fall into this category:

....

An alternative method of costing and charging is based on HOURLY RATE.

Once an hourly rate is established for repair work the labour charge for a job would be calculated as follows:

Complete the formula

Hourly rate ....

The hourly rate charge for work is the sum of:
1. Hourly wage rate of technician
2. Overhead percentage or establishment charge
3. Profit

The hourly wage rate of the technician is a DIRECT COST incurred on the job. OVERHEADS, which are INDIRECT COSTS, are all the costs involved in running the business.

Make a list of the items which fall into this category:

Rent, rates ....

Electricity, gas, water charges ....

Insurance charges ....

....

....

....

....

....

....

....

....

....

....

....

....

Fixed and Variable Costs

Some costs will vary according to the volume of business, others are fixed irrespective of the work load.

Give typical examples of:

Fixed Costs	Variable Costs
....	....
....	....

To determine the overhead costs for a particular department, the costs incurred directly by that department plus a proportion of the general overheads (rent etc.) are taken into account.

This example shows how hourly rate charged for repair work may be calculated.

Assume:

    Annual overhead costs are £600 000
    Annual wage costs are £200 000

Establishment charge $= \dfrac{600\,000}{200\,000} \times 100 = 300\%$

Technician wage rate per hour = £10.00

Establishment charge per hour = £10.00 × 300% = £30.00

A typical profit margin is 20% of turnover

Therefore hourly rate charged = £10.00 (wage) + £30.00 (overheads) + £10.00 (profit)
= £50.00/hour

NOTE: The cost of parts and materials would be added to the customer's bill.

CONTRACTS

A basic knowledge of the law relating to contract is essential for personnel dealing with customers, e.g. staff selling goods or services. Incorrect or misleading statements or omissions from them – written or spoken – made in connection with a transaction, can make the individual and/or the company liable to serious court action.

A CONTRACT is a legally binding agreement between parties and should either of the parties involved not keep to the bargain, the other can sue for breach of contract.

State the essential ingredients of a contract:

There must be an OFFER and an unqualified ACCEPTANCE.
....
....
....
....
....
....
....
....
....
....
....

State the meaning of the word CONSIDERATION in the context of a contract:

....
....
....
....

A SIMPLE CONTRACT is where both parties give and receive something of value. How does this differ from a SPECIALITY CONTRACT?

....
....
....
....

Terms of a Contract

The terms of a contract are the parts of the contract on which the parties have agreed, e.g. the service, goods, price, delivery, etc.

EXPRESS TERMS are terms which have been put into words orally or in writing. What are IMPLIED TERMS in a contract?

....
....
....
....
....
....
....

A contract may include a number of express or implied terms, some of which are more important than others. A CONDITION is a vital term in a contract:

Distinguish between a CONDITION and a WARRANTY:

....
....
....
....

Discharge of a Contract

When the parties to a contract have fulfilled all their contractual obligations the contract is DISCHARGED and is no longer binding or in force.

In what other ways can a contract be discharged?

1. *The parties can have a legally binding agreement to discharge the contract.*

2. ....
....

3. ....
....
....

What is an EXEMPTION or EXCLUSION clause in a contract?

....
....

Distinguish between VOID, VOIDABLE, ILLEGAL and UNENFORCEABLE contracts:

VOID   *This is where the contract is not legally enforceable.*

VOIDABLE   *This is when one of the parties can, at some stage, cancel*

               the contract.

ILLEGAL   ....
....

UNENFORCEABLE   ....

....
....
....
....

In a situation where goods are being sold, e.g. a car, a salesperson may make a false statement to enhance the goods and tempt the buyer to enter into a contract; this is known as ....

The remedies open to the buyer or innocent party when misrepresentation occurs are RESCISSION and ....
Rescission is when the contract is terminated and the parties to the contract are restored to their situation prior to the contract being formed.

What is meant by damages?

....

A contract can become void or voidable as a result of DURESS occurring while negotiating the contract. What is duress?

....
....
....
....

Give an example of a contract of the UTMOST GOOD FAITH and briefly describe DISCLOSURE in relation to this:

....
....
....
....
....
....
....
....
....
....

It is beyond the scope of this book to go into great detail regarding contract law. However, many books on the subject are available in libraries for further background reading. A typical car purchase contract is shown on the following pages. The layout and detailed wording of such documents may vary from firm to firm. They all form a binding legal agreement between the buyer and the seller.

Rogers

Rogers Corner, 274 Fylde Road, Preston PR2 2NJ
Tel: (0772) 722482 Parts: (0772) 735359
Fax: (0772) 722289

V.A.T. Reg. No. 364 0406 76

NEW/USED VEHICLE ORDER FORM
(Delete Where Applicable)

To..

I/WE HEREBY AGREE TO PURCHASE FROM YOU **SUBJECT TO THE TERMS AND CONDITIONS HEREOF (including those overleaf)** THE UNDERMENTIONED VEHICLE, EXTRAS AND ACCESSORIES, HEREINAFTER CALLED THE "GOODS"

(hereinafter described as the "Seller")

DETAILS OF VEHICLE BEING ORDERED

Make	Model	Colour	Trim
Chassis No./Frame No	Engine No	Key No.(s) Door	Stock No.
Reg. No.	Date First Reg.	Ignition	
		Odometer Reading	

Recorded mileage cannot be relied upon as the actual mileage run by the vehicle

✓ Where applicable

	New	Used
Car	Van	
Truck	Retail	
Fleet	Trade	
Motorcycle		

Customer setting balance by:-
☐ CASH ☐ CHEQUE

	£	p	%	£	p	VAT

VEHICLE—SPECIAL BODYWORK, FACTORY FITTED OPTIONS

ACCESSORIES TO BE FITTED

SUBJECT TO MANAGEMENT APPROVAL

SOLD

	@	VAT	@	VAT
Accessories				
Sub Total				
Car Tax				
Delivery				
Fuel .... Galls				
Number Plates.				
V.A.T.				
R. F. Licence .... months				
Insurance.				
Total Price				
Add H.P. Settlement to				
Sub Total				
Less P/Ex Allowance*				
Net Price				
Amount due				
from...................Finance Co.				
Balance				
Less Deposit				
Balance Due From Cust.				

TOTALS EXCL. VAT

£ _____ FORTHWITH AND TO PAY THE BALANCE AS SOON AS THE GOODS HAVE BEEN COMPLETED FOR DELIVERY AND NOTIFICATION THEREOF BEEN GIVEN TO ME/US BY POST AT THE ADDRESS BELOW, AND I CERTIFY THAT I AM 18 YEARS OF AGE OR OVER (where the Purchaser is an individual).

IN THE CASE OF PURCHASING A USED VEHICLE I/WE CERTIFY THAT BEFORE SIGNING THIS DOCUMENT MY/OUR ATTENTION HAS BEEN DRAWN TO THE AGE OF THE VEHICLE (AS SHOWN ABOVE) AND THE FACT THAT ANY DEFECTS MAY BE PRESENT ON THAT ACCOUNT. IN ADDITION I/WE UNDERSTAND THAT IT IS A TERM OF THE CONTRACT THAT I/WE SHOULD EXAMINE THE VEHICLE BEFORE SIGNING THIS ORDER FORM TO SATISFY MYSELF/OURSELVES AS TO ITS QUALITY AND THAT I/WE HAVE CARRIED OUT SUCH AN EXAMINATION, IN PARTICULAR MY/OUR ATTENTION HAS BEEN DRAWN TO THE FOLLOWING ITEMS: TYRES; BODY & PAINTWORK: GLASS; INTERIOR: TRIM & UPHOLSTERY: THE GENERAL CONDITION WITH RESPECT TO ITS AGE.

THIS DOCUMENT CONTAINS THE TERMS OF A CONTRACT. SIGN IT ONLY IF YOU WISH TO BE LEGALLY BOUND BY THEM. NOTHING HEREIN CONTAINED IS INTENDED, NOR WILL IT AFFECT, A CONSUMERS STATUTORY RIGHTS UNDER THE SALE OF GOODS ACT 1979 OR THE UNFAIR CONTRACT TERMS ACT 1977.

I/WE AGREE TO DEPOSIT THE SUM OF

"SELLERS" DECLARATION
I/WE ACCEPT AND CONFIRM THE ABOVE ORDER AND UNDERTAKE TO SUPPLY THE SAID GOODS UPON AND SUBJECT TO THE TERMS AND CONDITIONS REFERRED TO HEREIN

Signature...................................Date...............

Estimated Delivery Date................................

Place of Delivery................................

Salesman................................

DETAILS OF VEHICLE TO BE PURCHASED IN PART EXCHANGE

Make	Model	Registration No.	Colour	Date of last M.O.T. Test	Licence Expires
Engine No.	Chassis No./Frame No.	Date First Reg.	Odometer Reading	Date of Appraisal	Appraisal No.

*Allowance given in P/Ex for above: £ _____

Purchaser's Signature.................................

Name (Block Capitals)................................

Address................................

Tel No................................

THE VEHICLE IS/IS NOT SUBJECT TO ANY LIEN OR ENCUMBRANCE, IF IT IS STATE DETAILS THE ABOVE MILEAGE IS/IS NOT CORRECT. (IF NOT CORRECT THE APPROXIMATE TRUE MILEAGE IS:............miles)
IT WAS/WAS NOT PURCHASED BY ME NEW.
IT WAS/WAS NOT USED ABROAD BEFORE BEING REGISTERED IN THE U.K.
IT HAS/HAS NOT BEEN USED FOR SELF DRIVE HIRE, HACKNEY CARRIAGE OR TAXI WORK.
IT HAS/HAS NOT BEEN INVOLVED IN ANY ACCIDENT WHICH RESULTED IN A TOTAL LOSS CLAIM.
I HAVE READ AND UNDERSTAND THE SELLERS TERMS AND CONDITIONS AND VERIFY THE DETAILS SUPPLIED BY ME/US ARE CORRECT

Used Car Owner's Signature.....................................Date...............

THIS FORM HAS BEEN APPROVED BY THE MOTOR AGENTS' ASSOCIATION

TERMS AND CONDITIONS

Nothing herein contained is intended to affect, nor will it affect, a consumer's statutory rights under the Sale of Goods Act 1979 or the Unfair Contract Terms Act 1977.

1. This order and any allowance in respect of a used motor vehicle offered by the Purchaser are subject to acceptance and confirmation in writing by the Seller.

2. Any accessories fitted as new to the vehicle will be entitled to the benefit of any warranty given by the manufacturers of those accessories.

3. (a) The Seller will endeavour to secure delivery of the goods by the estimated delivery date (if any) but does not guarantee the time of delivery and shall not be liable for any damages or claims of any kind in respect of delay in delivery. (The Seller shall not be obliged to fulfil orders in the sequence in which they are placed.)

   (b) If the Seller shall fail to deliver the goods within 21 days of the estimated date of delivery stated in this contract the Purchaser may by notice in writing to the Seller require delivery of the goods within 7 days of receipt of such notice. If the goods shall not be delivered to the Purchaser within the said 7 days the contract may be cancelled.

4. If the contract be cancelled under the provisions of clause 3 hereof the deposit shall be returned to the Purchaser and the Seller shall be under no further liability.

5. If the Purchaser shall fail to take and pay for the goods within 14 days of notification that the goods have been completed for delivery, the Seller shall be at liberty to treat the contract as repudiated by the Purchaser and thereupon the deposit shall be forfeited without prejudice to the Seller's right to recover from the Purchaser by way of damages any loss or expense which the Seller may suffer or incur by reason of the Purchaser's default.

6. The goods shall remain the property of the Seller until the price has been discharged in full. A cheque given by the Purchaser in payment shall not be treated as a discharge until the same has been cleared.

7. Where the Seller agrees to allow part of the price of the goods to be discharged by the Purchaser delivering a used motor vehicle to the Seller, such allowance is hereby agreed to be given and received and such used vehicle is hereby agreed to be delivered and accepted, as part of the sale and purchase of the goods and upon the following further conditions:

   (a) (i) that such used vehicle is the absolute property of the Purchaser and is free from all encumbrances;

   or (ii) that such used vehicle is the subject of a hire purchase agreement or other encumbrance capable of cash settlement by the Seller, in which case the allowance shall be reduced by the amount required to be paid by the Seller in settlement thereof;

   (b) that if the Seller has examined the said used vehicle prior to his confirmation and acceptance of this order, the said used vehicle shall be delivered to him in the same condition as at the date of such examination (fair wear and tear excepted);

   (c) that such used vehicle shall be delivered to the Seller on or before delivery of the goods to be supplied by him hereunder, and the property in the said used vehicle shall thereupon pass to the Seller absolutely;

   (d) that without prejudice to (c) above such used vehicle shall be delivered to the Seller within 14 days of notification to the Purchaser that the goods to be supplied by the Seller have been completed for delivery;

   (e) that if the goods to be delivered by the Seller through no default on the part of the Seller shall not be delivered to the Purchaser within 30 days after the date of this order or the estimated delivery date, where that is later, the allowance on the said used vehicle shall be subject to reduction by an amount not exceeding 2¹/₂% for each completed period of 30 days from the date of the expiry of the first mentioned 30 days, to the date of delivery to the Purchaser of the goods.

8. In the event of the non-fulfilment of any of the foregoing conditions, other than (e) the Seller shall be discharged from any obligation to accept the said used vehicle or to make any allowance in respect thereof, and the Purchaser shall discharge in cash the full price of the goods to be supplied by the Seller.

9. Notwithstanding the provisions of this agreement the Purchaser shall be at liberty before the expiry of 7 days after notification to him that the goods have been completed for delivery to arrange for a finance company to purchase the goods from the Seller at the price payable hereunder. Upon the purchase of the goods by such finance company, the preceding clauses of this agreemtn shall cease to have effect, but any used vehicle for which an allowance was thereunder agreed to be made to the Purchaser shall be bought by the Seller at a price equal to such allowance, upon the conditions set forth in clause 7 above (save that in (c), (d) and (e) thereof all references to "delivery" or "delivered" in relation to "the goods" shall be construed as meaning delivery or delivered by the Seller to or to the order of the finance company) and the Seller shall be accountable to the finance company on behalf of the Purchaser for the said price and any deposit paid by him under this agreement.

10. If the goods to be supplied by the Seller are new, the following provisions shall have effect:

    (a) this agreement and the delivery of the goods shall be subject to any terms and conditions which the Manufacturer or Concessionaire may from time to time lawfully attach to the supply of the goods or the re-sale of such goods by the Seller, and the Seller shall not be liable for any failure to deliver the goods occasioned by his inability to obtain them from the Manufacturer or Concessionaire or by his compliance with such terms or conditions. A copy of the terms and conditions currently so attached may be inspected at the Seller's Office;

    (b) the Seller undertakes that he will ensure that the pre-delivery work specified by the Manufacturer or Concessionaire is performed and that he will use his best endeavours to obtain for the Purchaser from the Manufacturer or Concessionaire the benefit of any warranty or guarantee given by him to the Seller or to the Purchaser in respect of the goods and, save in the case of consumer sales (as defined by the Sale of Goods (Implied Terms) Act 1979) all statements, conditions or warranties as to the quality of the goods or their fitness for any particular purpose whether express or implied by law or otherwise are hereby expressly excluded;

    (c) notwithstanding the sum for Car Tax specified in the order, the sum payable by the purchaser in respect thereof shall be such sum as the Seller has legally had to pay or becomes legally bound to pay for Car Tax in respect of the goods and notwithstanding also the sum for Value Added Tax specified in the order, the sum payable by the Purchaser in respect thereof shall be such sum as the Seller becomes legally liable for at the time the taxable supply occurs;

    (d) if after the date of this order and before delivery of the goods to the Purchaser the Manufacturer's or Concessionaire's recommended price for any of the goods shall be altered, the Seller shall give notice of any such alteration to the Purchaser, and

       (i) in the event of the Manufacturer's or Concessionaire's recommended price for the goods being increased the amount of such increase which the Seller intends to pass to the Purchaser shall be notified to the Purchaser. The Purchaser shall have the right to cancel the contract within 14 days of the receipt of such notice. If the Purchaser does not give such notice as aforesaid the increase in price shall be added to and become part of the contract price;

       (ii) in the event of the recommended price being reduced the amount of such reduction, if any, which the Seller intends to allow to the Purchaser shall be notified to the Purchaser. If the amount allowed is not the same as the reduction of the recommended price the Purchaser shall have the right to cancel the contract within 14 days of the receipt of such notice;

    (e) in the event of the Manufacturer of the goods described in the order ceasing to make goods of that type, the Seller may (whether the estimated delivery date has arrived or not) by notice in writing to the Purchaser, cancel the contract.

11. (a) If a used vehicle is supplied as roadworthy at the date of delivery and, in the case of consumer sales (as defined by the Supply of Goods (Implied Terms) Act 1979):

    (i) is sold subject to any conditions or warranties that are implied by the Sale of Goods Act 1979 or any amending statute;

    (ii) prior to signing this order form the Purchaser shall examine the vehicle and the items set out in the Purchaser's Certificate of Examination overleaf and, the Purchaser is reminded that the condition of merchantable quality implied by Section 14(2) of the Sale of Goods Act 1979 does not operate in relation to such defects which that examination ought to reveal. Should the goods be sold also subject to defects notified by the dealer to the Purchaser before signing the agreement, the condition of merchantable quality above referred to does not operate in relation to those defects.

    (b) Save in the case of consumer sales (as defined) all statements, conditions or warranties as to the quality of the goods or their fitness for any purpose whether expressed or implied by law or otherwise are hereby expressly exluded.

SERVICING RECORDS

Service records provide:

(a) *a definition of the work to be carried out.* ....

(b) *a record of the work carried out.* ....

(c) *a record of time spent.* ....

(d) ....

(e) ....

(f) ....

(g) ....

(h) ....

(i) ....

When a vehicle is booked in at a garage for a service or repair job, the 'system of information' relating to the job varies considerably owing to the size and nature of motor-trade establishments.

In a modern service complex, a computer network with terminals in reception, parts department, car sales and accounts is utilised to provide all the relevant information. A typical procedure is outlined opposite.

In addition to a history of work done on a customer's vehicle, what other information is included on a customer record?

....

....

....

....

In the space opposite describe briefly a system of job control with which you are familiar.

The typical system described is for general guidance only.

(1) Customer and reception agree on work required and customer signs job order, a copy of which he or she retains.
(2) Details of job are entered on to the workshop loading sheet for the job, date and time.
(3) Details of customer, vehicle and job are fed into reception computer and a job card print out, with job number is obtained for the workshop. (Job cards are desirably prepared in the afternoon prior to the day on which the jobs are started.)
(4) On completion of the job the technician enters labour time on to the job card which is fed into reception computer, and this, together with materials consumed (this information is in the computer network from the parts terminal) is used to prepare customer's invoice on reception computer.
(5) Customer service record is updated on reception computer.

....

....

....

....

....

....

....

....

....

....

....

....

....

....

....

PARTS DEPARTMENT

The parts department offers a spares service to the public and provides the parts required by the workshop within a service complex. Its location within the complex is important.

State the factors that should be taken into account when planning its location:

Access for vehicles loading and unloading stock.
....

....

....

....

....

Make a simple line diagram to show a plan view of a garage to which you have access. Show the location of the parts department relative to the other departments.

Storage

The type of storage equipment used depends upon the size, shape and weight of the parts being stored.
A type of storage unit used for storing small and medium sized parts is shown at 'A' opposite.
Sketch another type of storage unit at 'B' and state the purpose for which it would normally be used.

A   B

Purpose ....
....

Sketch in the space below the arrangement of storage units and issue points for a typical parts store.

Use the feint squares as a guide.

Stores Procedure

List the main activities directly related to the functioning of a parts department:

Inspection and receipt of stock
....

Stock control
....

....

....

....

Stock Control

Stock control is a management system to ensure:

1. *Parts in demand are always available but not overstocked.*
....

2. ....

Each part stocked has a record (normally on computer) bearing details of the part and a record of sales.
By keeping such a record, the stock order and hence stock levels can be related to demand.
What information would be on record for a particular component?

1. *Description of part (name and part number)*
....

2. *Number in stock.*
....

3. ....

4. ....

5. ....

6. ....

Identification and Withdrawal

When a request is made for a part, the parts person must first IDENTIFY the part and then LOCATE it.

Describe briefly a system used to identify a part, from what may be in certain instances quite a vague request from a customer.

....

....

....

....

....

....

Describe a method of part location:

....

....

....

Stock Taking

A complete physical stock check is normally carried out once or twice per year. The reasons for this are:

1. *For accounting purposes (annual accounts)*
....

2. ....

Describe a procedure for carrying out a physical stock check:

....

....

....

....

....

....

....

....

....

Chapter 3

Road Vehicle Systems and Layouts

VEHICLE BODIES – PURPOSE AND FUNCTIONAL REQUIREMENTS

The main purpose of the vehicle body is to accommodate driver, passengers and luggage, or in the case of a goods vehicle to carry various forms of goods, e.g. solids, liquids, bulky, heavy, etc.

Most car bodies are made from STEEL. Name two other materials used for car bodywork.

List other important factors relating to car body design and structure.

....

....

....

....

....

....

The car body must also provide mounting points for the main mechanical and electrical components, e.g. engine, transmission, suspension, steering, exhaust systems, wiring and lighting, seat belts and seating.

In recent years designers have paid considerable attention to the aerodynamic aspect of body shape. Why is this?

....

....

....

GOODS VEHICLE BODIES

Many goods vehicles consist basically of:

A CHASSIS frame on to which all the running gear, engine and transmission are mounted.

A CAB which provides comfortable, safe accommodation for driver and passengers.

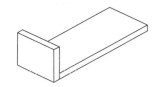

A BODY which is designed according to the type of loads to be carried.

In a RIGID type of goods vehicle the CAB and BODY are mounted on the same CHASSIS.

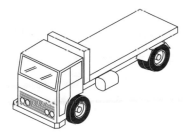

With an ARTICULATED type of goods vehicle the BODY is mounted on a separate SEMI-TRAILER chassis.

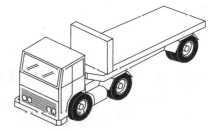

TYPES OF VEHICLE BODY

A vehicle body is designed according to the purpose for which the vehicle is intended.

Name the types of vehicle body shown opposite. Complete the table below to give examples of current vehicles and their respective body types.

Type	Make	Model
Saloon		

Name the type of vehicle shown below and list some of its features which make it very popular.

....

....

....

....

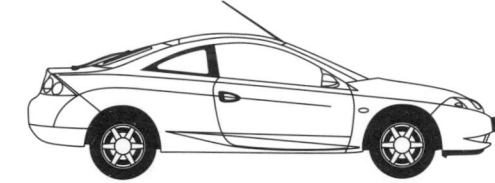

....

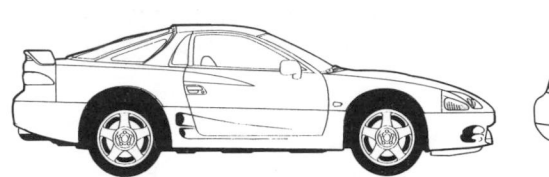

....

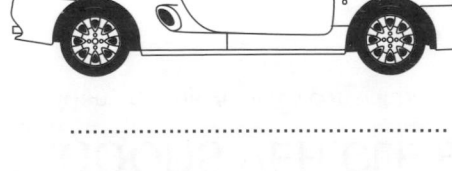

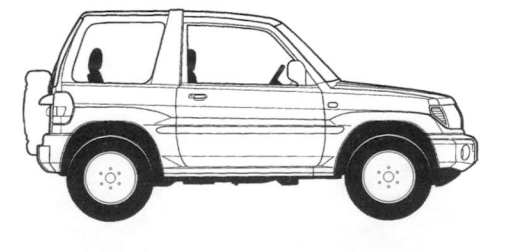

....

42

The main body types for goods vehicles are:

....

....

....

....

....

A popular type of light goods vehicle based on a car layout is illustrated below. Name this type of vehicle and give a reason for using it as opposed to a van or estate derivative of a car.

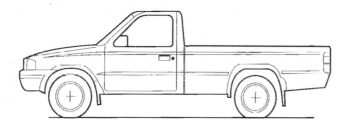

Type ....

Typical make .... Model ....

Reason for use

....

....

Add a TRUCKMAN TOP to the vehicle above.

Many vans have a combined (one piece) cab and body mounted on a separate chassis ('A' below); small vans are often of integral construction with body and floor (i.e. 'chassis') in one unit. Complete the illustrations at B and C by sketching on the chassis shown, the types of van body named.

A

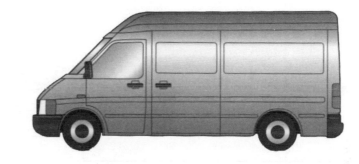

B

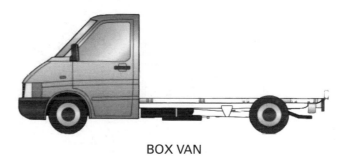

BOX VAN

C

LUTON VAN

Name the types of bodies shown below and give one advantage of each.

....

....

On some integral van bodies there is no 'bulkhead' between the payload area and the forward driving compartment. The driver can therefore step from the driving seat into the rear load space and then out through the nearside door to make a delivery. What is the name given to this type of van?

....

HEAVY GOODS VEHICLES (HGVs)

The main difference between HGVs and light goods vehicles is that the HGV, apart from being longer and heavier, usually has more than two axles.

Name the types of HGV below and add the labelling to the drawing at (c).

(a)

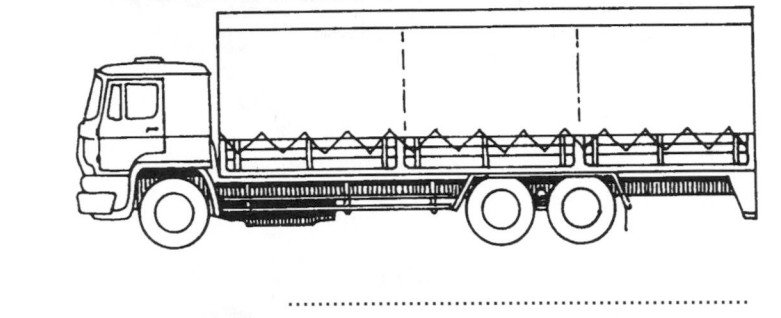

....

(b)

....

(c)

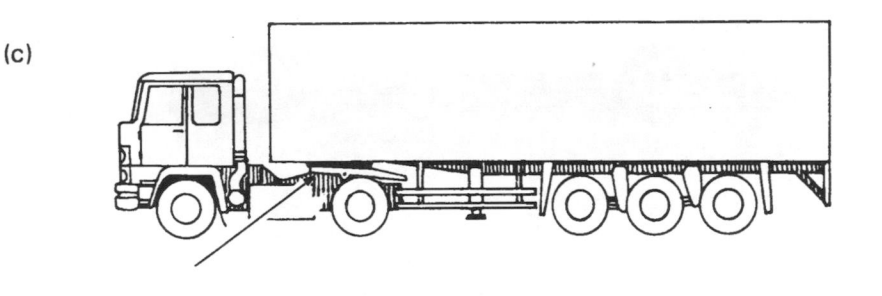

....

....

The fifth wheel coupling is the connecting mechanism for the semi-trailer. During cornering or manoeuvring, the semi-trailer pivots (articulates) about the fifth wheel coupling.

44

VEHICLE LAYOUTS

In the goods vehicle shown at (a) below, the driver sits forward of the front steered wheels. With the vehicle layout at (b) the driver sits behind the front steered wheels. Name both types of vehicle classification.

(a)

Vehicle type ....

(b)

Vehicle type ....

PASSENGER TRANSPORT VEHICLES

As with goods vehicles, passenger vehicle bodies are designed and constructed according to the type of work in which the vehicle will be engaged.

A coach body is shown below.

How does this body differ from the omnibus body as used for in-town multi-stop work?

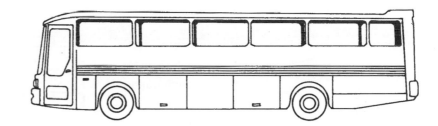

....

....

....

....

....

....

....

Name the type of passenger vehicle shown below.

....

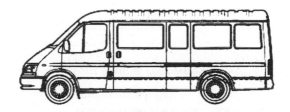

ENGINE LOCATIONS

State where on the following vehicles the engine position is likely to be. State its probable type, Spark Ignition (SI) or Compression Ignition (CI), and number of cylinders.

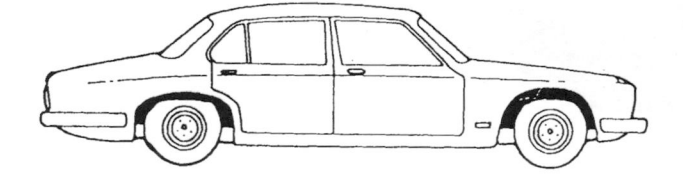

....

....

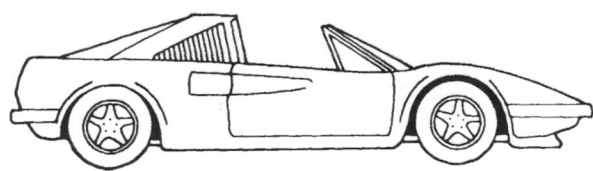

....

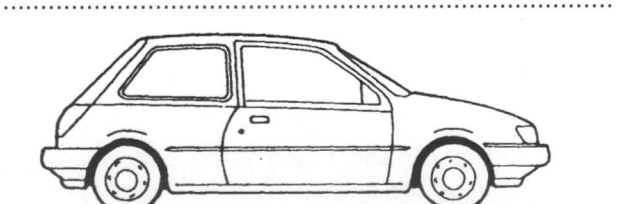

....

....

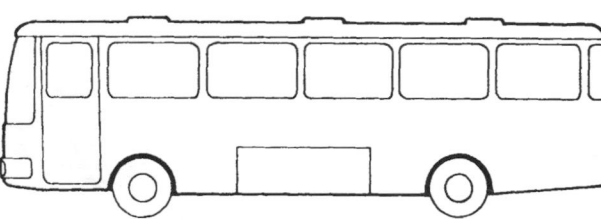

....

....

Examine manufacturers' literature and state typical vehicles which have

Front engine, rear wheel drive:

MAKE	MODEL
....	....
....	....
....	....

Front engine, front wheel drive:
(longitudinal)

MAKE	MODEL
....	....
....	....
....	....

Front engine, front wheel drive:
(transverse)

MAKE	MODEL
....	....
....	....
....	....

Rear engine:

MAKE	MODEL
....	....
....	....

Mid engine:

MAKE	MODEL
....	....

Underslung engine:

MAKE	MODEL

VEHICLE MECHANICAL LAYOUTS

The 'layout' of the motor vehicle is concerned with the arrangement of main mechanical components.

List the names of the main components:

1. .... 2. ....

3. .... 4. ....

   5. ....

Three layouts are shown opposite. Give a name to each layout and label the drawings.

State TWO advantages of layout A.

1. ....

2. ....

State TWO advantages of layout B.

1 ....

2. ....

Engines can be mounted longitudinally or transversely.

State TWO advantages of layout C.

1. ....

2. ....

State TWO disadvantages of a transversely mounted engine over a longitudinally mounted engine.

1. ....

2. ....

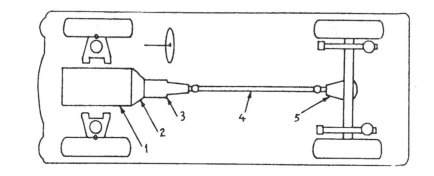

A

.... layout

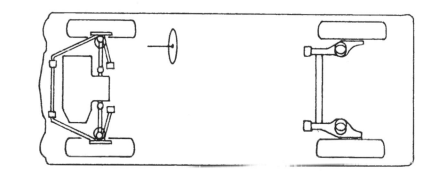

B

.... layout

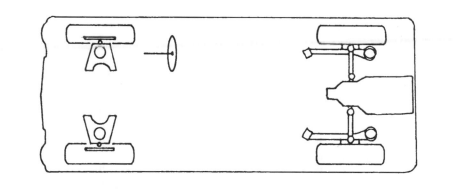

C

.... layout

In the sports car shown below, the engine is transversely mounted immediately behind the seating compartment driving the rear wheels. Many high-performance cars and almost all single-seat racing cars employ this arrangement, although the engine in single-seat racing cars is usually mounted longitudinally. Name this arrangement.

....

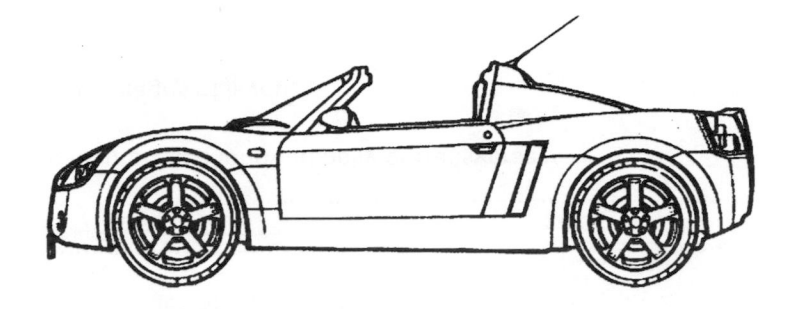

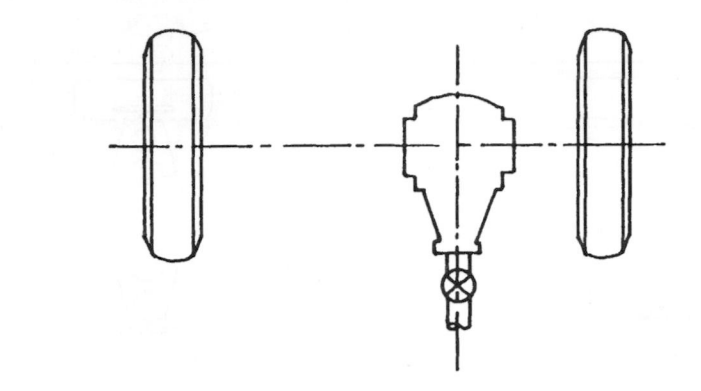

FOUR WHEEL DRIVE

As the name implies, four wheel drive (4WD) or (4 × 4) is where all the four road wheels are driven. In the past 4WD was used mainly on vehicles operating off the normal roads, on uneven, slippery or soft surfaces.

Two examples of such use are:

1. ....

2. ....

In recent years, owing to the refinements in 4WD systems, many new cars employ the system. It offers improved traction and control. Complete the layout opposite to show the main components in a 4WD system.

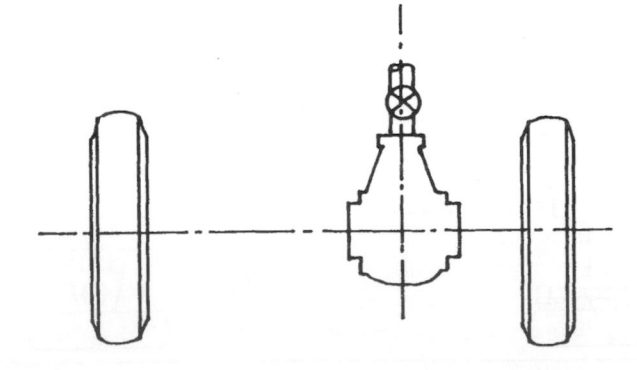

CHASSIS

The chassis is the framework of a vehicle on to which all the main components of a vehicle are attached. It may be a relatively heavy rigid structure which is separate from the body of the vehicle (as in goods vehicles) or, as illustrated below, a single body/chassis structure as adopted for most cars.

Name and label the structure below and explain briefly how rigidity and strength are achieved within the structure. Name and state the purpose of A and B.

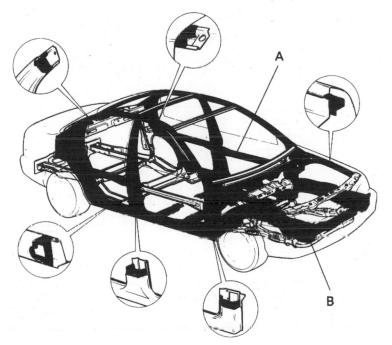

....
....
....
....
....
....

Make a sketch below to illustrate a separate chassis frame as used on a heavy goods vehicle. Indicate on the drawing where the main components, i.e. engine, transmission and suspension, would be attached to the frame.

Ladder Frame Type Chassis

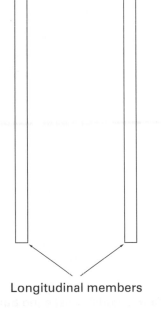

Longitudinal members

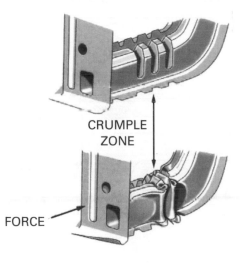

CRUMPLE ZONE

FORCE

The above drawings show bumper brackets with built-in crumple zones. Draw a circle around the areas which have helped to absorb the shock of an impact.

JACKING POINTS

The jacking points are the places on a vehicle body at which the jack is applied in order to lift a wheel (or wheels) clear of the ground. The body structure at the jacking points is made sufficiently strong to withstand the concentrated load imposed when the vehicle is jacked up.

How many jacking points are usually built into the vehicle body?

....

....

Investigation

Examine a vehicle and complete the drawing below to show the location of the jacking points. Also show by simple line outlines the position of the engine and transmission.

Make .... Model ....

Give two reasons why a vehicle's own jacking points are not normally used during servicing by a garage.

1. ....

....

2. ....

....

State the precautions to be observed when carrying out a jacking procedure using a vehicle side jack:

....

....

....

....

....

....

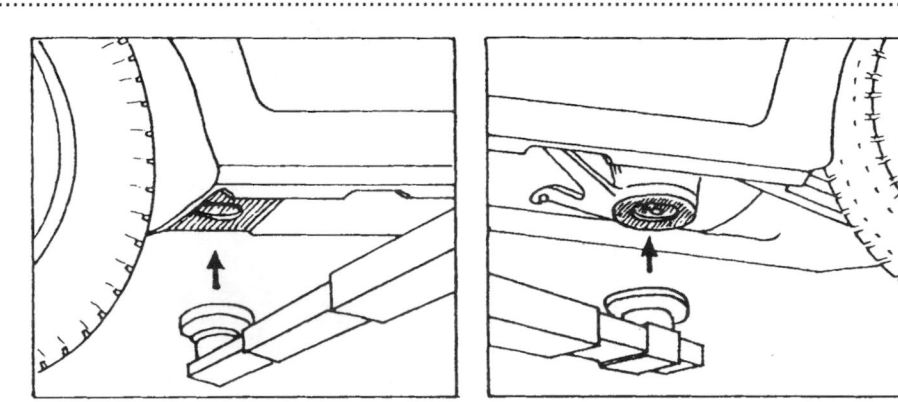

What do the drawings above illustrate?

....

....

....

....

....

....

....

....

SUBFRAMES, SOUND INSULATION AND MOUNTINGS

On a car the major components such as engine, transmission and suspension are sometimes not directly attached to the main integral body/chassis. The loads and stresses imposed on the car body by such components can be somewhat isolated from it by attaching the components to a rigid frame known as

a ....

A typical front subframe assembly is shown below with the various mounting points lettered, which are

(a) ....

(b) ....

(c) ....

(d) ....

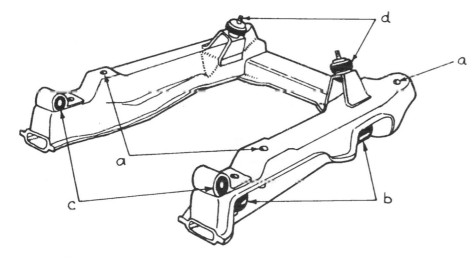

Sound Insulation

The body drawing (above right) illustrates one type of sound insulation and typical places where it is applied. State what material is likely to be used and its intended effect.

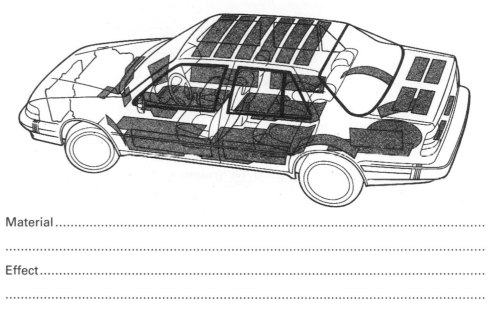

Material ....
....

Effect ....
....

Note: Plastic foam, thick felt or similar may be used additionally to dampen out resonances in body cavities and the like.

Flexible Mountings

The major units on a vehicle are usually attached to the chassis or body structure through flexible rubber mountings. Name and label the mountings shown which are for engine/transmission and suspension. Give reasons for using these mountings.

....
....
....

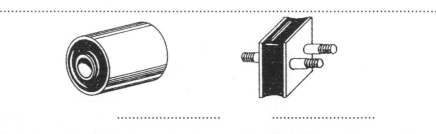

....

CAB MOUNTING

The cab of a goods vehicle is mounted on the chassis via flexible or sprung mountings. It may be fixed permanently in position or, as is the case with many forward control vehicles, the cab is hinged at the front to allow it to tilt when required. A fixed rubber-type cab mounting and a modern rubber-spring, leading arm, front tilt cab mounting are shown below; complete the labelling on the drawings.

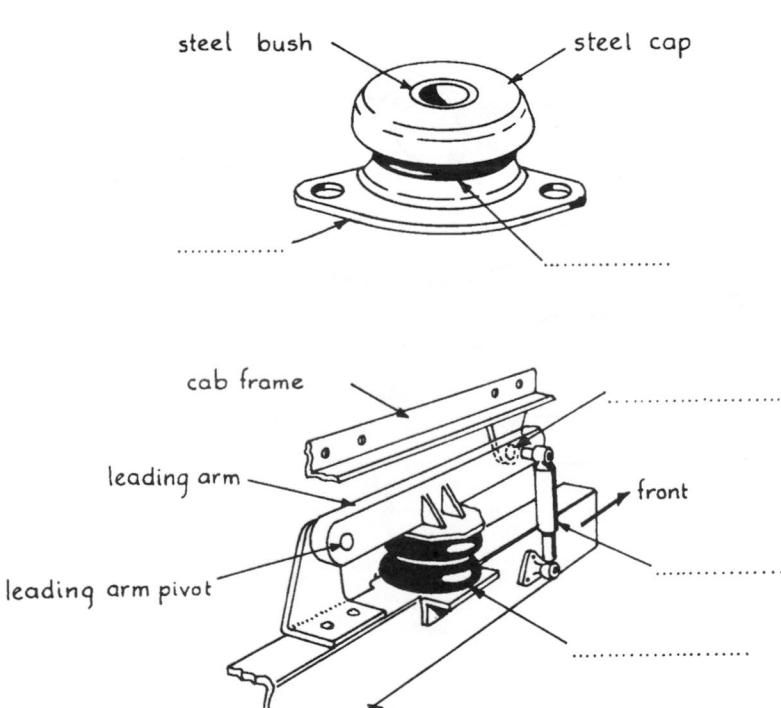

steel bush   steel cap

....

cab frame   ....

leading arm   front

leading arm pivot   ....

....

chassis frame

What benefits are to be obtained by allowing a cab to tilt forward?

....

....

....

A modern tilt cab mounting arrangement is shown below. The system acts rather like a vehicle suspension system using springs and dampers to isolate the driver further from road shocks and vibration.

Label the drawing.

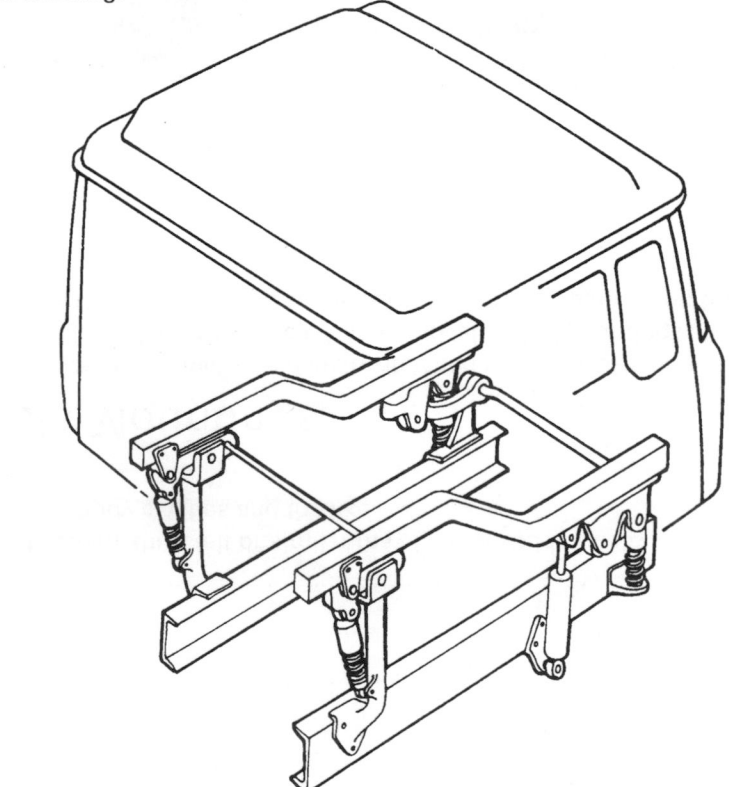

Describe the operational procedure to tilt the cab.

....

....

....

....

LOAD PLATFORM ATTACHMENT

One method of securing the load platform or body of a goods vehicle to the chassis is shown at (a) below. Examine a goods vehicle similar to the one shown below and sketch an alternative method of body attachment at (b). Indicate on the vehicle drawing below the number and position of these attachments.

(a)   **(b)**

U bolt ——→ body frame

wood or plastic strip

chassis frame

plate ——→

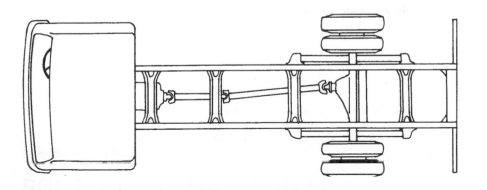

The doors on most cars and goods vehicles are hinged at their forward ends. The van shown below has a side sliding door.

Give reasons for using this arrangement in preference to hinged doors on many vehicles of this type:

1. ....

....

2. ....

....

....

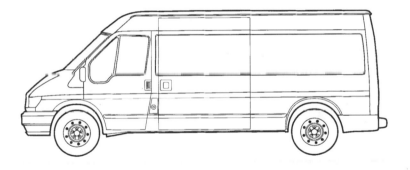

Give examples of places on a vehicle body where access HATCHES and FLAPS are used.

HATCHES..

FLAPS ....

BODY (OCCUPANT SAFETY)

Seat Belts

The restraining force required to prevent car occupants from being thrown forward when a collision occurs is extremely high. Seat belt anchorage points on the vehicle body must therefore be sufficiently strong to withstand the loads involved.

The seat belt is normally wound on a spring loaded reel when it is not in use. An amount of belt is drawn off the reel according to the seat position and the size of the wearer when it is secured in place.

It allows the wearer freedom of movement when belted in and locks into position in the event of a sudden stop.

It is known as the   REEL system.

Add arrows to the drawing to indicate the seat belt anchorage points.

The illustration shows the action of a device which is part of the seat belt mechanism.

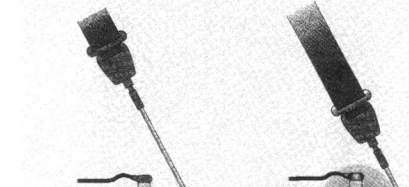

Name the device and state its purpose.

....

....

....

....

Airbags – Secondary Restraint System (SR)

In the event of a severe frontal impact above 15 to 18 mph, a driver's airbag will spontaneously inflate to reduce head and chest injuries which even a properly restrained driver can suffer. Airbags can also be provided for the front seat passengers and for side impact.

The system is shown below: name parts 1,2,3 and 4.

1. ....

2. ....

3. ....

4. ....

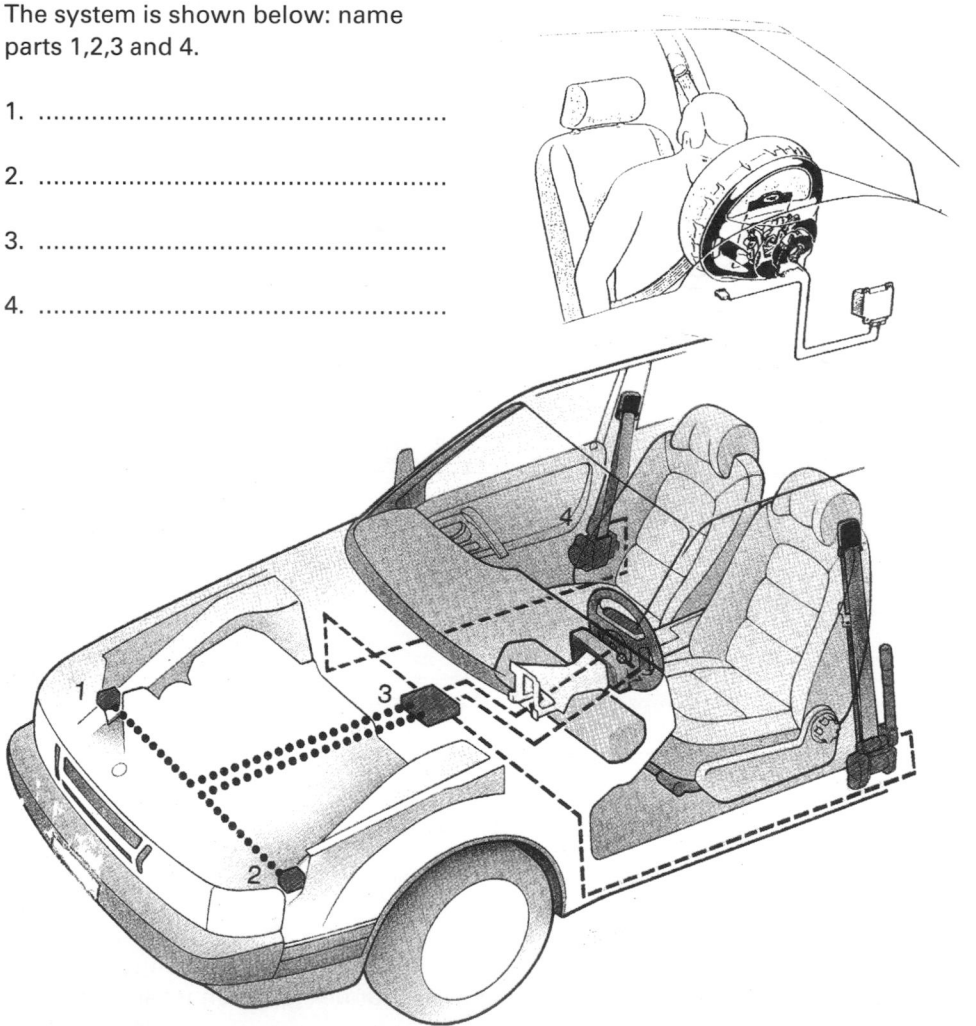

Body Exterior

The illustration below shows some of the panels which are attached to the basic integral structure of a car.

Name the panels shown.

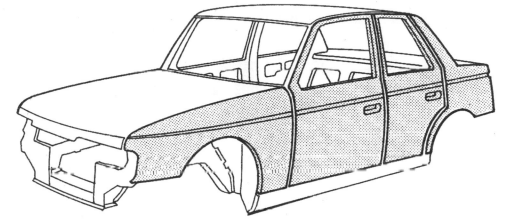

Name three methods used to attach these panels to the vehicle:

....

....

....

External Trim

The external trim on modern cars is usually made up of a number of plastic mouldings or chromium-plated strips which are bolted or clipped on to the body panels.

Investigation

Examine a modern car and name the parts that constitute external trim.

Car make .... Model ....

Component	Secured by
....	....
....	....
....	....
....	....
....	....

A typical bumper attachment (Rover) is shown below. State briefly how it is attached:

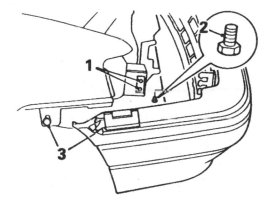

....

....

....

....

....

....

....

....

....

Vehicle Windows

The windows on a vehicle may be HINGED, DROP, FIXED or SLIDING. Give examples of places on a vehicle where each type of window is used.

HINGED ....

DROP ....

FIXED ....

SLIDING ....

The illustration below shows the weather sealing arrangement of a car door. Examine a car boot and sketch the sealing arrangement employed.

Door Seal

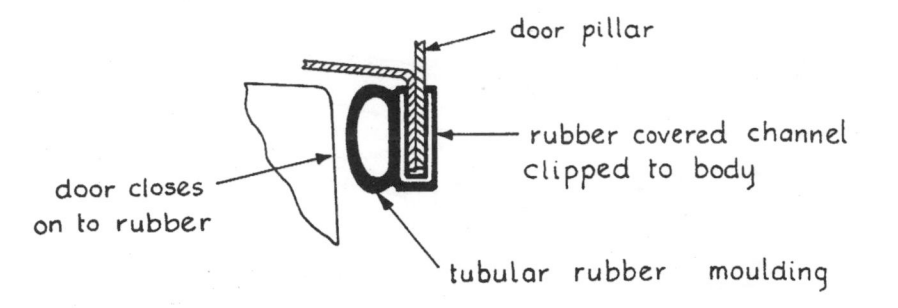

door pillar

rubber covered channel clipped to body

door closes on to rubber

tubular rubber moulding

Boot Seal

Interior Trim

The interior trim of a vehicle can be extremely basic, as on lower priced models, or quite lavish as on some luxury models.

Name some of the items which make up the interior trim on a car:

....

....

....

....

Seats

The illustration shows the construction of a car seat. Label the drawing and state the purpose of part (A).

....

....

....

....

....

....

....

....

A →

FOAM RUBBER

COVER

HEATED PANELS

LUMBAR SUPPORT ADJUSTER

BACKREST RAKE ADJUSTMENT

HEIGHT ADJUSTER

FORE & AFT ADJUSTMENT

ENGINES

The engine of a motor vehicle provides the POWER to propel the vehicle and to operate the various ancillaries, such as the alternator, water pump, vacuum-assisted brakes, air conditioning and such like. However, when the accelerator is released and the road wheels are made to turn the engine, it provides a useful amount of vehicle RETARDATION without the use of the brakes.

ENGINE TYPES

Engines used in motor vehicles may be referred to as FOUR-STROKE or TWO-STROKE engines.

The term stroke refers to ....

....

A four-stroke engine is one in which ....

....

A two-stroke engine is one in which ....

....

The number of revolutions completed during a working cycle on a four-stroke engine is ....

The number of revolutions completed during a working cycle on a two-stroke engine is ....

Investigation

Complete the tables to identify different types of engines used in motor vehicles.

FOUR-STROKE ENGINES

Vehicle	No. of cylinders	Type of fuel	Engine capacity

TWO-STROKE ENGINES

Vehicle	No. of cylinders	Type of fuel	Engine capacity

Identify the engines shown in terms of two and four stroke.

State two reasons why each engine can be so identified.

Name the arrowed parts.

Type ....

Reason for identification

(a) ....

....

....

(b) ....

....

....

Type..

Reason for identification

(a) ....

....

....

(b) ....

....

....

ENGINE – FUNCTIONAL REQUIREMENTS

The law of conservation of energy states that energy cannot be destroyed, it can only change from one form to another

In what way does an engine receive energy and into what form of energy is it converted?

....

....

....

What is the basic function of an engine?

....

....

List the major vehicle systems that are combined, interrelated or are interactive with the engine:

1. ....

2. ....

3. ....

4. ....

5. ....

6. ....

List the major components that make up the systems listed above:

1. ....

....

....

2. ....

....

....

3. ....

....

....

4. ....

....

....

5. ....

....

....

6. ....

....

....

Identify on the sketch below the major systems that are combined, interrelated or interactive with the engine.

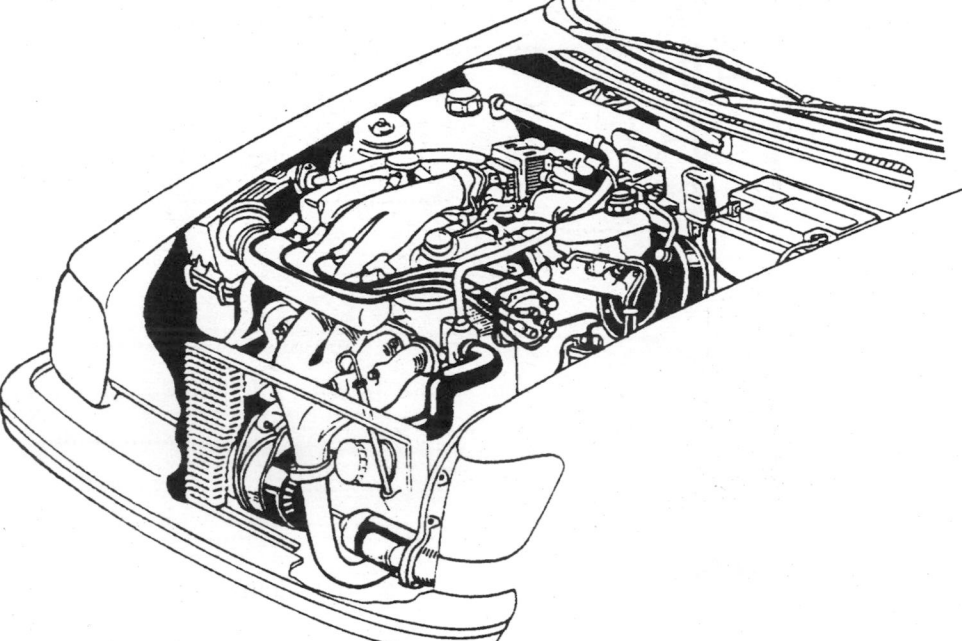

ENGINE CONSTRUCTIONAL FEATURES

Name the parts indicated on the engine shown opposite:

1. ....
2. ....
3. ....
4. ....
5. ....
6. ....
7. ....
8. ....
9. ....
10. ....
11. ....
12. ....
13. ....
14. ....
15. ....
16. ....
17. ....
18. ....
19. ....
20. ....
21. ....
22. ....
23. ....
24. ....
25. ....
26. ....

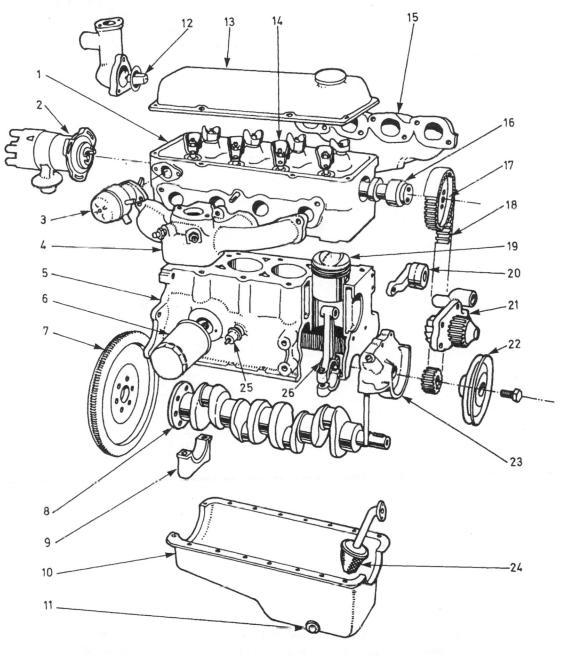

THE OTTO (OR FOUR-STROKE) CYCLE

The four-stroke cycle is completed in four movements of the piston during which the crankshaft rotates twice.

Complete the line diagrams to show the positions of the valves at the commencement of each stroke. Indicate the direction in which the piston is moving in each case.

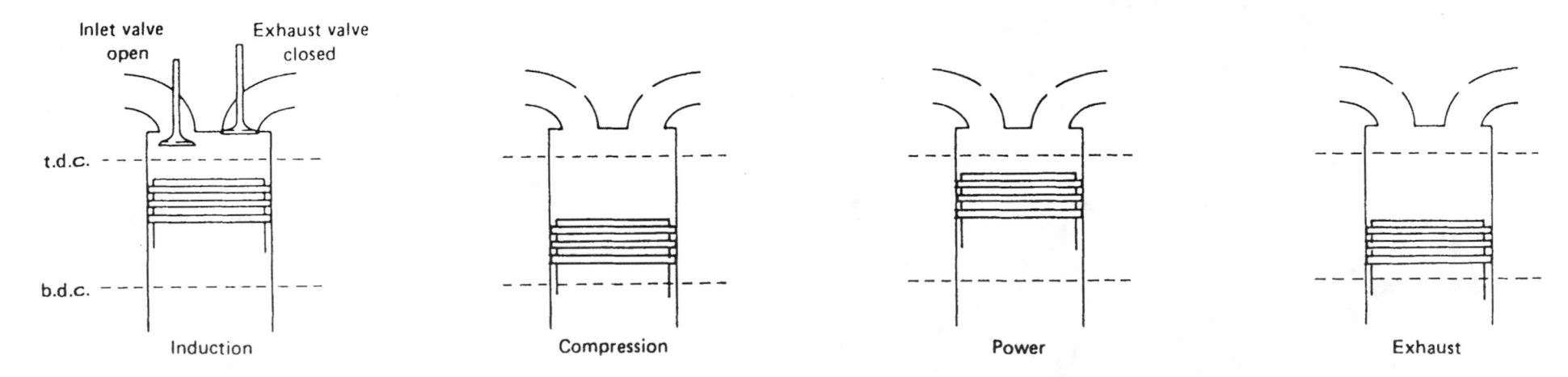

Induction   Compression   Power   Exhaust

Investigation

By use of a sectioned four-stroke spark-ignition engine, note the sequence of operations of the piston, valves and spark when the engine is rotated, and describe what happens on each stroke when the engine is running.

Induction	Compression	Power	Exhaust
....	....	....	....
....	....	....	....
....	....	....	....
....	....	....	....
....	....	....	....
....	....	....	....
....	....	....	....
....	....	....	....

TWO-STROKE PETROL ENGINE CYCLE (CRANKCASE COMPRESSION TYPE)

By making use of both sides of the piston, the four phases, induction, compression, power, exhaust, are completed in two strokes of the piston or one crankshaft revolution.

Mixture entering and leaving the combustion chamber is controlled by the piston. It acts as a valve covering and uncovering ports in the cylinder wall.
Induction into the crankcase can be controlled by the piston (via the piston port) by reed valves or by a rotary valve.

Explain why mixture enters the crankcase.

....
....
....
....
....

Explain why the mixture transfers from the crankcase to the combustion chamber.

....
....
....
....
....

Explain how a two-stroke engine is lubricated.

....
....
....
....

Name the main parts and show the direction of crankshaft rotation.

(a) Show fuel entering crankcase.

(b) Show fuel transferring to cylinder and exhausting.

(c) Indicate reed valve on alternative intake system.

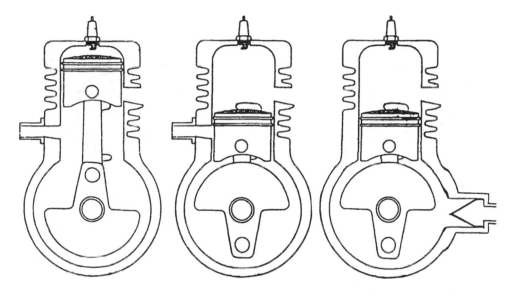

Summary of two-stroke cycle

Stroke	Upward	Downward
Events above piston	Closing of transfer port Completion of exhaust Compression	Expansion of gases Commencement of exhaust Transfer of mixture from below piston
Events below piston	Induction of new mixture into crankcase	Partial compression of new mixture in crankcase

COMPRESSION-IGNITION (DIESEL) ENGINE – FOUR-STROKE CYCLE

The actual strokes, induction, compression, power and exhaust, are exactly the same as in the spark-ignition engine. However the operating principle is slightly different, as described below.

INDUCTION   The piston is descending and with the inlet valve open air only enters the cylinder.

COMPRESSION   When the inlet valve closes the air is compressed to between 16:1 and 24:1. This causes the air temperature to rise substantially. Approximately 25° b.t.d.c. fuel is injected into the cylinder in the form of a fine spray; it mixes rapidly with the air and the high temperature causes combustion to take place.

POWER   Combustion causes a rapid pressure rise and the piston is forced downwards.

EXHAUST   The ascending piston forces the burnt gases out of the cylinder.

List the main differences of the compression-ignition engine when compared with the spark-ignition engine.

....

....

....

....

....

....

....

....

....

....

....

Examine and compare the basic components of a spark-ignition and compression-ignition engine of similar size.

Component	Ways in which compression-ignition engine components differ from typical spark-ignition engine components
Piston	....
Connecting rod	....
Crankshaft	....
Cylinder block	....

Combustion Chamber Shapes – Diesel

Describe the combustion chamber shapes used in diesel engines.

direct injection

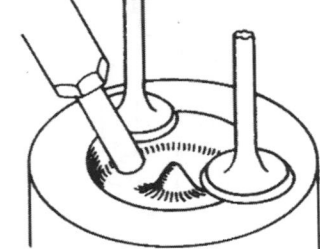

indirect injection

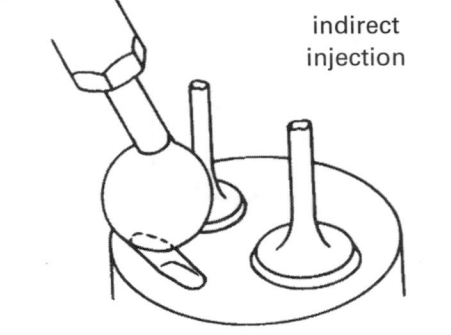

....

....

....

....

Compression-Ignition Engine Advantages

The compression-ignition engine has certain advantages when compared with the spark-ignition. List the main advantages.

....

....

....

TWO-STROKE CYCLE – C.I. (DIESEL) ENGINE

As with the four-stroke compression-ignition cycle, initially air only enters the cylinder.

In order to complete the cycle in two strokes the air must be forced into the cylinder by means of a pressure charger. It is usual that air induction is through ports and exhaust by poppet valves.

Describe the operational cycle of the engine shown.

....

....

....

....

....

....

....

....

....

....

The method of arranging the valves and ports is shown below.

Name the important parts and show, using arrows, the air flow through the system.

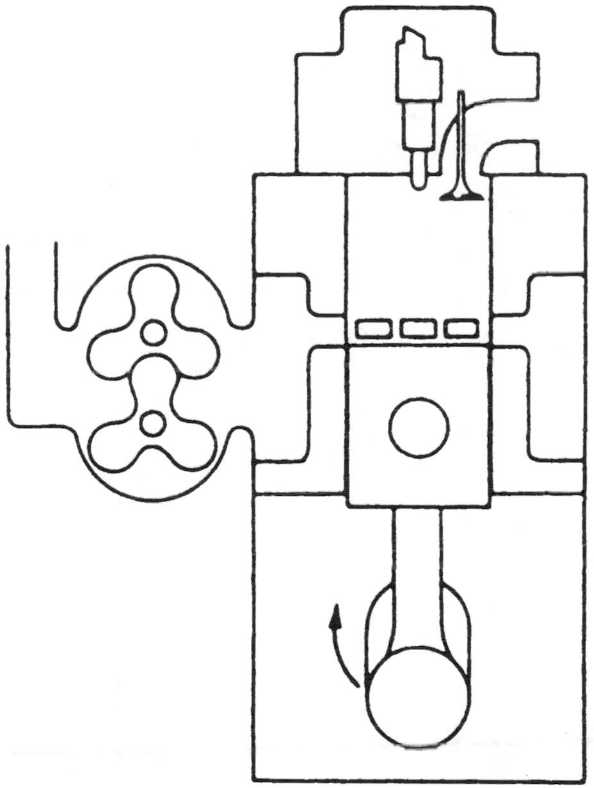

Uniflow two-stroke compression-ignition engine

63

TYPES OF COMBUSTION CHAMBERS USED IN SPARK-IGNITION ENGINES

State the features that make these types of combustion chambers suitable for spark-ignition engines and how they help to minimise toxic exhaust emissions.

Bath tub

....
....
....
....
....
....
....
....
....
....

Wedge

....
....
....
....
....
....
....
....
....
....

Hemispherical

....
....
....
....
....
....
....
....
....
....

Bowl in piston

....
....
....
....
....
....
....
....
....
....

CYLINDER VOLUME

The volume of an engine cylinder is found by multiplying the area of the cylinder end by the distance moved by the piston (stroke).

Area of cylinder end $= \dfrac{\pi d^2}{4}$ or πr^2

where d = cylinder bore

Swept volume $=$

Insert the appropriate dimension abbreviations on the drawing. For example, calculate the swept volume of an engine cylinder having a bore diameter of 84 mm and a stroke of 90 mm.

Take π as $\dfrac{22}{7}$

(Since engine capacity is quoted in cubic centimetres (cm³), sometimes written as cc, the basic dimensions should be first converted to cm.)

Therefore bore 84 mm = 8.4 cm and r = 4.2 cm
stroke 90 mm = 9.0 cm

Swept volume = $\pi r^2 \times$ stroke

$= \dfrac{22}{7} \times 4.2 \times 4.2 \times 9$

$= 22 \times 0.6 \times 4.2 \times 9$

$= 498.96$ cm³

If this cylinder was from a four-cylinder engine it would be a .... litre engine.

1. Calculate the swept volume of an engine cylinder having a bore diameter of 70 mm and a stroke of 100 mm.

....
....
....
....
....
....
....
....

2. Calculate the swept volume of an engine cylinder having a bore diameter of 80 mm and a stroke of 70 mm.

....
....
....
....
....
....
....

3. Calculate the capacity of a four-cylinder engine whose bore and stroke are both 90 mm.

....
....
....
....
....
....
....

4. What is the volume of fuel contained in a cylindrical fuel tank 600 mm diameter and 900 mm long, when (a) it is completely full and, (b) two-thirds of the fuel has been used?

....
....
....
....
....
....
....

ENGINE CONSTRUCTIONAL FEATURES

The exploded view shows most of the main engine structure and moving parts used in a four-cylinder in-line engine with overhead camshaft.

Observe how the components would fit together to form the complete engine.

Study the interrelationship of all the moving parts, how each component is dependent on others for its operation and how they contribute to the overall design.

It is expected that you can identify
the main components on these
drawings from previous experience.

The numbered items are small but
essential parts of the engine.

Name these components.

No.	Name
1.	....
2.	....
3.	....
4.	....
5.	....
6.	....
7.	....
8.	....
9.	....
10.	....
11.	....
12.	....
13.	....
14.	....
15.	....
16.	....
17.	....
18.	....
19.	....
20	....

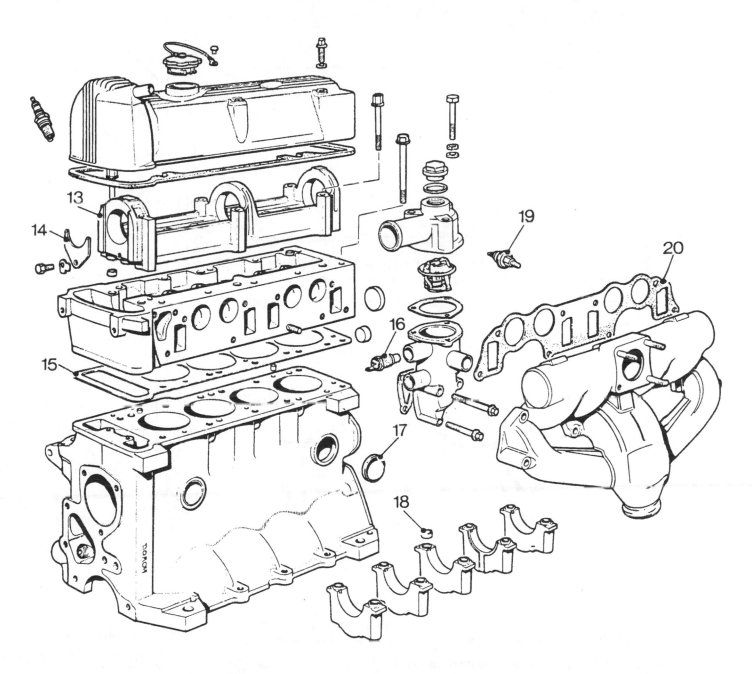

VALVE OPERATION

One type of valve arrangement is the overhead valve (ohv), with side camshaft.

Name the main parts
which operate the valve.

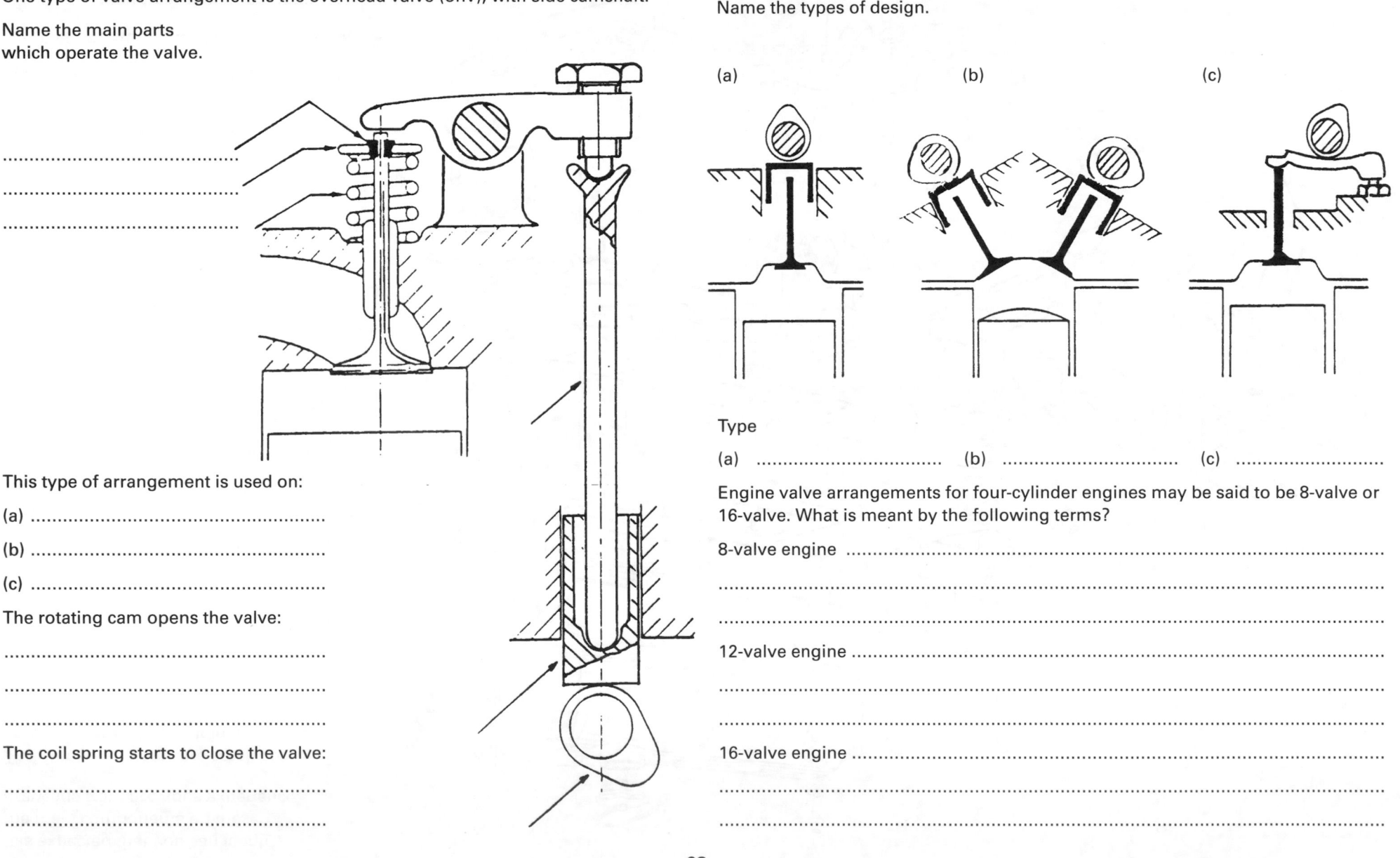

....

....

....

This type of arrangement is used on:

(a) ....

(b) ....

(c) ....

The rotating cam opens the valve:

....

....

....

The coil spring starts to close the valve:

....

....

Other valve arrangements are used on overhead camshaft (ohc) engines.

The sketches below show three different types of ohc valve layout.
Name the types of design.

(a)   (b)   (c)

Type

(a) .... (b) .... (c) ....

Engine valve arrangements for four-cylinder engines may be said to be 8-valve or 16-valve. What is meant by the following terms?

8-valve engine ....

....

....

12-valve engine ....

....

....

16-valve engine ....

....

ENGINE LUBRICATION SYSTEM

Name the parts indicated, which make up the engine lubrication system.

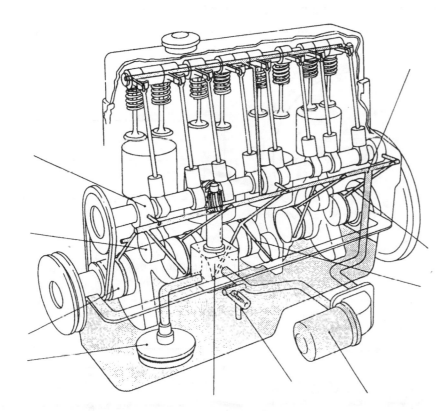

State FOUR functions provided by the oil:

....

....

Name a vehicle using this type of engine: ....

State the purpose of the following items shown opposite:

Gauze filter ....

....

Oil pump ....

....

Relief valve ....

....

External (secondary) filter ....

....

Main gallery ....

....

Investigation

Examine an engine so as to determine the position of the oil passage-ways, filters, pump and pressure relief valves etc.

Engine make .... Model ....

Component	Type	Position
Oil pump		
Primary filter		
Secondary filter		
Oil pressure relief valve		
Oil pressure indicator		

Reduction of Friction by Lubrication

When examined under a microscope, even apparently flat, smooth surfaces look like mountain ranges. For example, this even applies to a newly ground big end journal and its new shell bearings. The sketch below shows how two surfaces interlock with one another when they are dry.

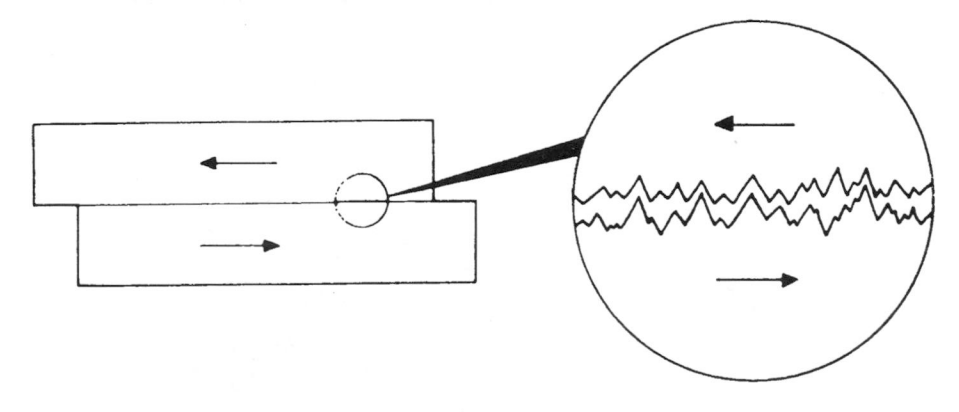

Complete the sketch to show the effect of lubrication.

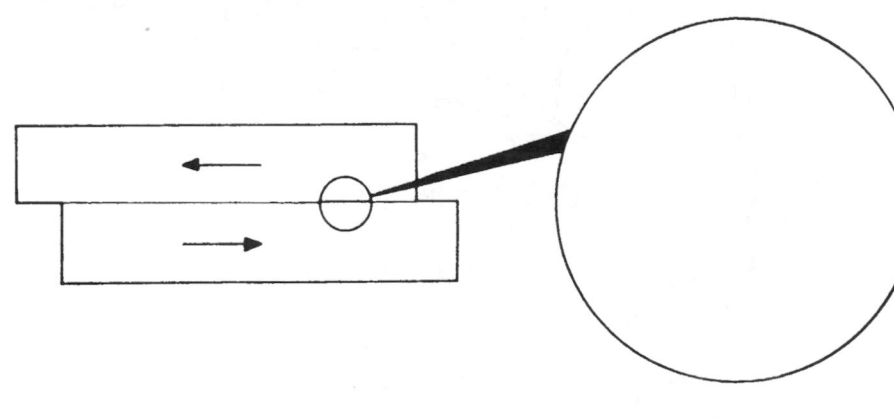

The effects of the lubricant are to ....

....

....

....

Friction can be an advantage or a disadvantage depending upon the situation. For example, it is essential to prevent the driver's foot slipping off the brake pedal, but is highly undesirable in engine bearings.

Name FOUR vehicle parts where friction is an:

Advantage	Disadvantage
1.	1. ....
2.	2. ....
3.	3. ....
4.	4. ....

Bearings

There are two main types of bearing:

(a) plain bearings in which .... is produced by the relative movement of the bearing surfaces.

(b) ball or roller bearings in which .... is produced by the relative movement of the bearing surfaces.

Which type of friction, 'sliding' or 'rolling', offers the least resistance to motion?

....

Viscosity

The viscosity of an oil is a measure of its thickness (or body), or (more correctly) of its ....

The viscosity of an oil is expressed as a number prefixed by the letters SAE, for example SAE 30. SAE stands for the Society of Automotive Engineers which is the American organisation that devised these viscosity standards.

State TWO typical engine oil viscosity numbers:

1 ....   2 ....

Note: the SAE number signifies the viscosity of the oil at the specified temperature. It in no way indicates the quality of the oil.

COOLING SYSTEMS

All types of internal combustion engines require some form of cooling.

Why is a cooling system necessary?

....

....

....

....

Any cooling system must be designed to prevent both under and over cooling.

Air Cooling

This type of cooling is popular for motor cycles and is used on some older small cars. It is also widely used on small portable engines which drive pumps and generators.

On what basic principle does air cooling rely?

....

....

....

....

....

Complete the table in respect of a variety of vehicles having air-cooled engines.

Vehicle make	Model	Number of cylinders	Approximate engine capacity

A motor-cycle engine is not usually enclosed; the passage of air over the cylinders when the cycle is in motion provides an adequate cooling flow.

Two examples of motor-cycle engine barrels and cylinder heads are shown. State the engine cycle on which each example would work.

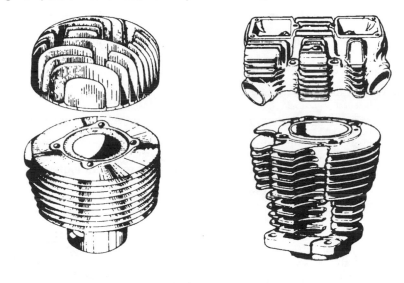

....

What is the reason for the variation in the size and shape of the cooling fins?

....

....

....

....

....

What periodic cooling-system maintenance should be carried out on air-cooled engines?

....

....

Pump-assisted Water Cooling System

On the diagram of the cooling system below, indicate the direction of water flow and identify the main components. Complete the sketch to show where the expansion tank would be fitted.

Show the recommended initial water level.

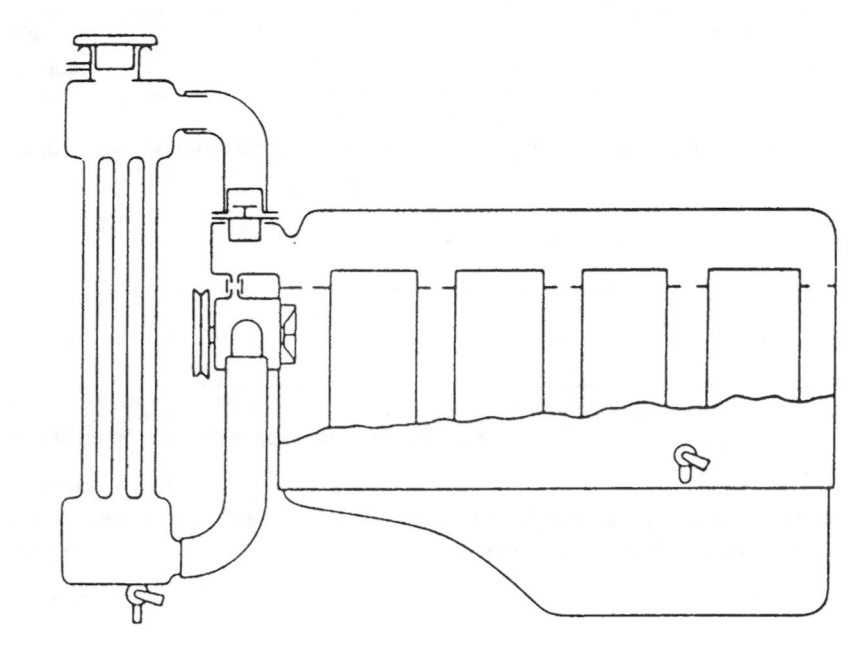

EXPANSION TANKS

All modern cooling systems use an expansion tank. The layout will then be called, depending upon type, a sealed or semi-sealed system.

What is the function of the expansion tank?

....

....

....

When an engine is started from cold, limited coolant circulation occurs through a by-pass hose or passage within the engine. There is no circulation into the radiator. The by-pass system prevents local overheating. As the engine reaches its normal operating temperature, the thermostat opens allowing coolant to flow through the radiator. The coolant cools as it passes through the radiator core and returns into the engine cool.

What item controls the rate of water flow?

....

What item controls the temperature at which the water boils?

....

Identify the cooling system components shown below.

 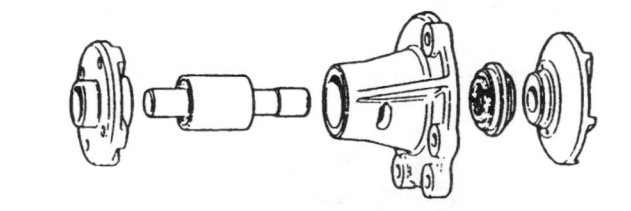

....

What are four advantages of air cooling over water cooling?

1. ....

2. ....

3. ....

4. ....

What are four advantages of water cooling over air cooling?

1. ....

2. ....

3. ....

4. ....

HEAT TRANSFER

Three methods by which heat is transferred are:

1. From the hotter to the colder part of a material by travelling within the actual material; this is known as ....

2. Through the atmosphere by means of rays; this is known as ....

3. By the movement of liquids or gases; this is known as ....

In many instances heat is dissipated from the heat source by more than one method at a time.

How is the heat dissipated from the friction surfaces in a drum brake assembly?
....
....
....

Heat will flow in one direction only ....
....

Give examples of materials which are

Good conductors of heat

....
....
....
....

Poor conductors of heat

....
....
....
....

Conduction

Pistons (and some cylinder heads and blocks) are made of aluminium alloy, instead of cast iron or steel. There are two reasons for the choice of this material. One is that it is lighter, the other is
....

This leads to the fact that the rate of conduction varies
....
....

Convection

Convection currents occur because of the difference in density between cold and hot water. In practical terms water density varies little through its complete liquid state, but compared to warm water the density of cold water is
....

What occurs to a quantity of water as its temperature rises? ....
....
....

The changes in density therefore directly produce ....
....

Radiation

Metal will lose heat at different rates according to:

1. ....
....

2. ....
....

3. ....
....

PRESSURE

An excellent example of pressure and change in pressure is in an engine cylinder. The mixture, which is a form of gas is drawn into the cylinder and as the piston

rises, the pressure is ....

The mixture is then ignited and the pressure is further ....

In the first case a reduction in.. caused the pressure to

.... while in the second case an increase in

.... caused the .... in pressure.

The intensity of pressure may be defined as 'the applied force per unit area acting at right angles'.

or Pressure = _____

The SI unit of pressure is the Pascal (Pa).

1 Pa =   ,   1 kPa =   ,   1 MPa =   or

When dealing with forces and areas, it is more usual to use the force/area units and then convert them to Pascals.

Examples

1. Calculate the pressure acting on a piston when a total force of 16 kN is applied to the piston crown whose cross-sectional area is 0.004 m^2.

   ....
   ....
   ....
   ....
   ....
   ....

2. Calculate the pressure in a brake cylinder when a force of 450 N is applied to the piston whose cross-sectional area is 150 mm^2.

   ....
   ....
   ....
   ....
   ....
   ....

Investigation

To show that gases exert pressure and are compressible.

1. Use a pump or compressed air to pressurise the container. Note the reading. ....
   Release the pressure.

2. Stand apparatus in hot water or in water being heated.

Results and observations

Pressure reading when cold ....

Pressure reading when hot ....

Conclusion ....

....

3. Calculate the pressure acting on a radiator cap when the spring force is 24 N and the cap area resisting force is 6 cm^2 (0.0006 m^2).

   ....
   ....
   ....
   ....
   ....
   ....

4. Calculate the force acting on a piston when the average pressure is 900 kPa and the cross-sectional area of the piston is 0.008 m^2.

   ....
   ....
   ....
   ....
   ....
   ....

LAYOUT AND PURPOSE OF MAIN PARTS IN A PETROL FUEL SYSTEM

Name the main parts of the fuel system layout shown:

....

....

State the purpose of each part of the fuel system listed below and comment upon their construction or design:

Fuel tank ....
....

Tank unit ....
....

Pipe line ....
....

Filters ....
....

Lift pump ....
....

Carburettor or injection unit ....
....

Air cleaner ....
....

Examine a fuel system layout and complete the table.

Make of vehicle.................................... Model ....

Component	Type, position or material where applicable
Air cleaner or silencer	
Carburettor or injection unit	
Filter	
Fuel pump	
Fuel pipes	
Fuel tank	

Single Point Injector Throttle Body

Name the numbered parts indicated

1. ....

2. ....

3. ....

4. ....

5. ....

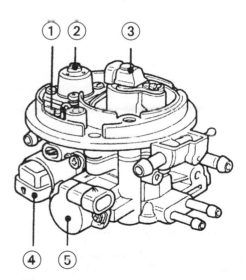

The Simple Carburettor

A simple single-jet, fixed-choke carburettor has all the main features of a modern sophisticated carburettor.

Label the major parts of the simple carburettor shown below.

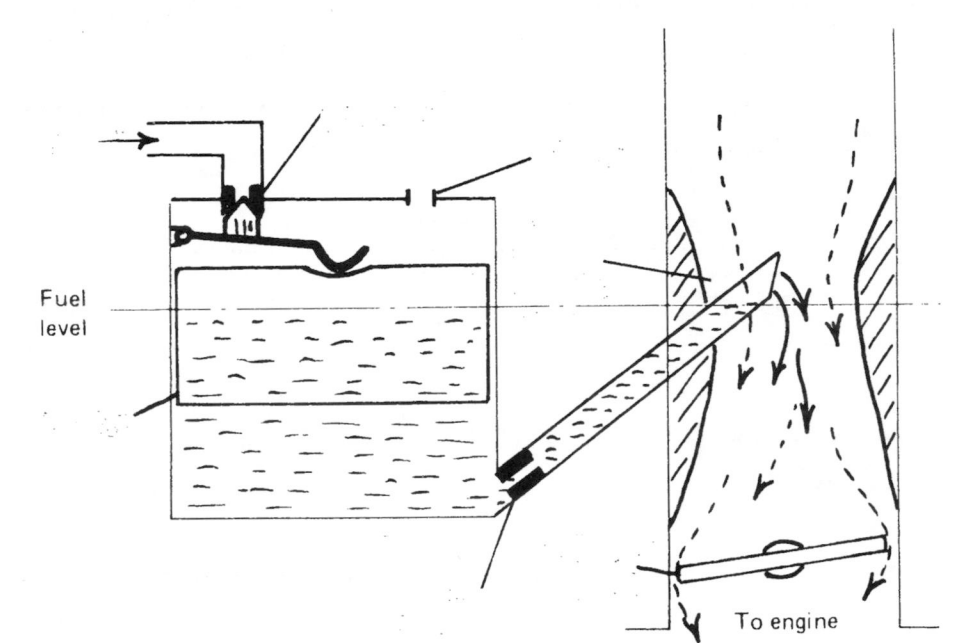

Fuel level

To engine

A simple carburettor's basic operating system is described below, add the missing words.

Fuel is maintained at the level shown. When the engine starts up the air passing

through the .... speeds up and reduces the .... pressure at

this point. The fuel in the ...., which is maintained

at atmospheric pressure, is forced through the .... into the

.... where it mixes (or atomises) into a fine vapour with the air and

passes to the engine.

What is the function of the venturi or choke tube?

....

....

What is the function of the float and needle valve?

....

....

A simple single-jet carburettor is not suitable for use on a modern variable speed engine. Why?

....

....

....

The graph below shows what would happen if a simple single-jet carburettor was fitted to a variable-speed engine.

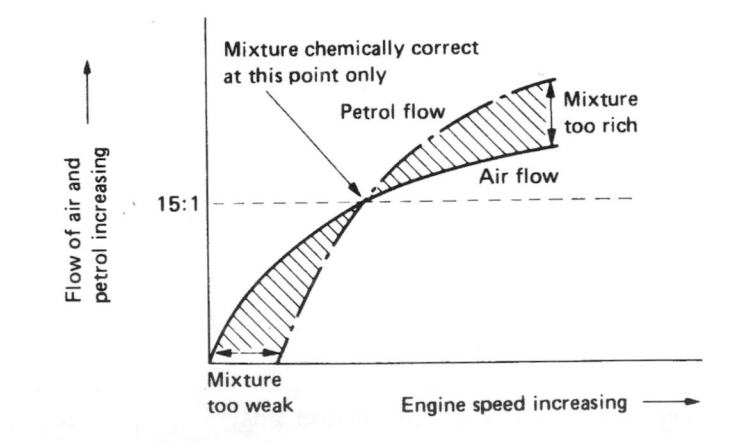

At low speeds the mixture strength would be ....

At high speeds the mixture strength would be ....

Constant Choke Carburettor

Name the basic parts of the constant choke carburettor shown below.

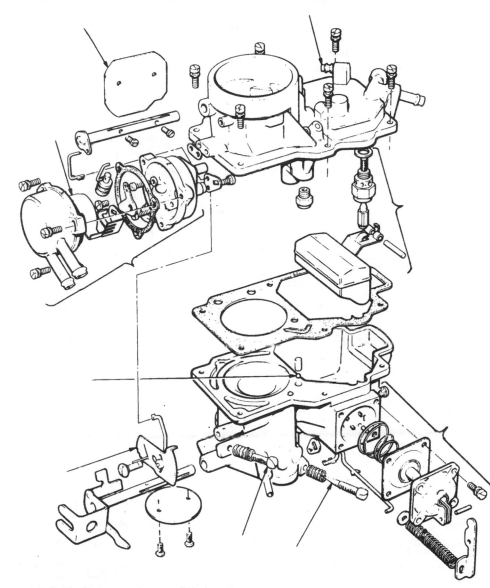

Constant Depression Carburettor

The drawings show the basic construction of a constant depression (variable choke) carburettor.

Carburettor
shown in
idling
position

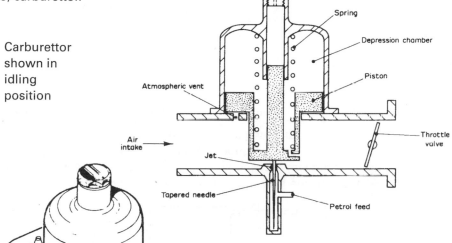

Spring

Depression chamber

Atmospheric vent

Piston

Air intake

Jet

Throttle valve

Tapered needle

Petrol feed

Complete the drawing below by showing the air flow through the carburettor and indicate how it lifts the piston.

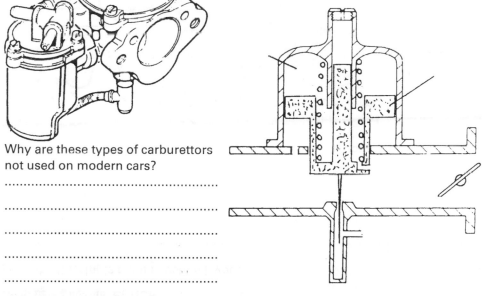

Why are these types of carburettors not used on modern cars?

....
....
....
....
....

FUEL SYSTEM LAYOUT – PETROL INJECTION

There are TWO different ways in which injectors are commonly positioned in the inlet manifold. These positions are shown on the two diagrams on this page.

MULTI POINT INJECTION

Describe the multi point injection layout:

....

....

....

....

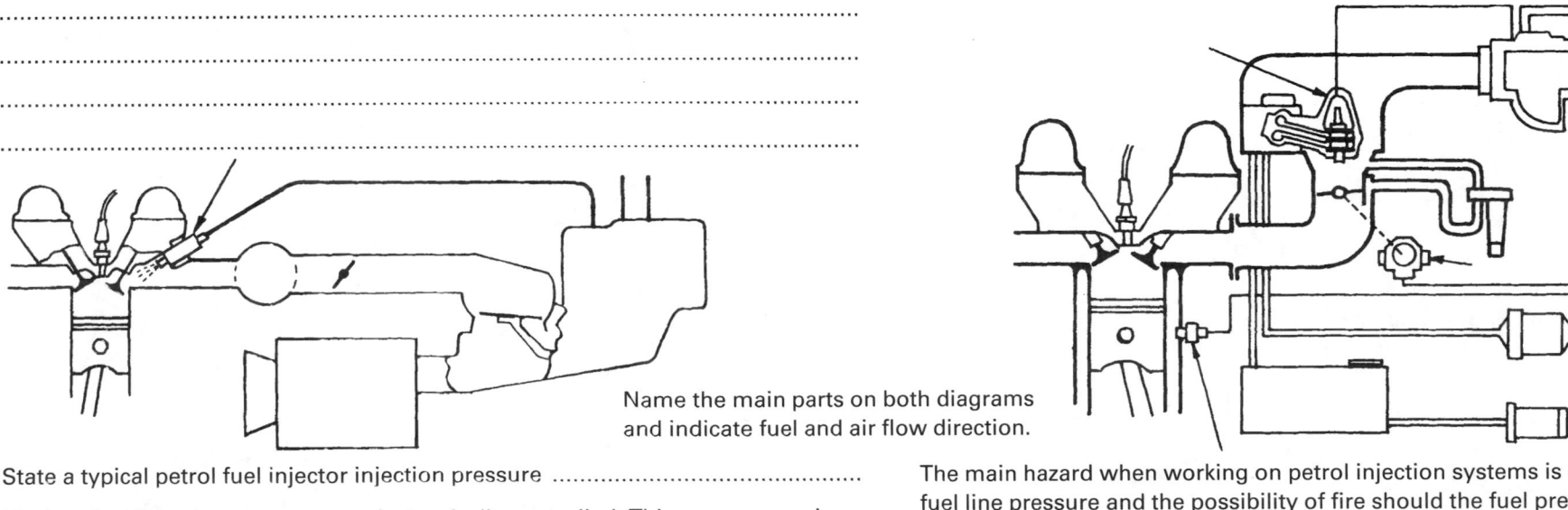

Name the main parts on both diagrams and indicate fuel and air flow direction.

State a typical petrol fuel injector injection pressure ....

Modern fuel injection systems are electronically controlled. This ensures precise fuel metering at all engine speeds. When compared with a carburettor this advantage has the effect of:

1. ....
2. ....
3. ....
4. ....

What are the most common problems that cause apparent failure in electronically controlled systems used on modern vehicles?

1. ....
2. ....
3. ....

SINGLE POINT INJECTION

Describe the single point injection layout:

....

....

....

....

The main hazard when working on petrol injection systems is created by the high fuel line pressure and the possibility of fire should the fuel pressure be suddenly released and the fuel be allowed to spray about.
Note. Fuel can self-ignite if sprayed on to a hot handlamp bulb.

Describe the precautions to take when, for example, changing a fuel filter:

1. ....
2. ....
3. ....
4. ....
5. ....
6. ....

FUEL SYSTEM LAYOUT – COMPRESSION IGNITION (DIESEL) ENGINE

ROTARY PUMP

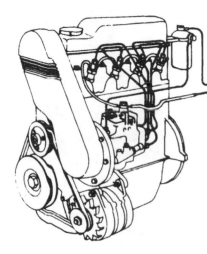

On what type of vehicle or engine size is the rotary pump layout most commonly used?

....
....
....
....
....
....
....

INLINE PUMP

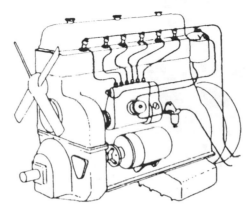

On what type of vehicle or engine size is the inline pump layout most commonly used?

....
....
....
....
....
....
....

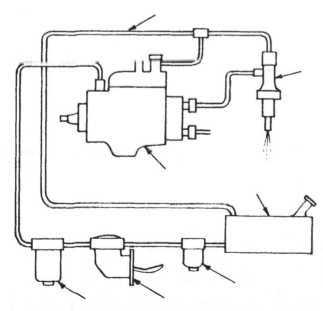

On both drawings indicate the direction of fuel flow and name the main parts.

Why is a two-stage fuel filtration system necessary?

....
....
....
....
....
....

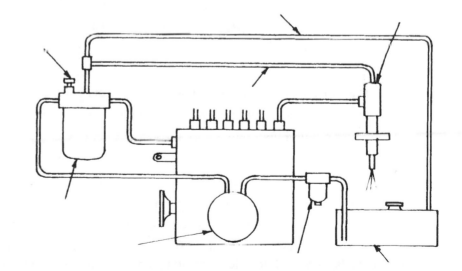

Common Rail Injection System

Diesel engined cars and light vans now tend to use direct (rather than indirect) injection systems. This design change on small diesel engines has basically been made possible by the use of electronically controlled fuel injectors, it's effect reduces emission levels and maintains fuel economy.

The common rail injection system is a high pressure direct injection (HDI) system. Its main advantage is that the injection pressure does not vary as engine speed is increased. This allows the engine to perform consistently well over its total operating range.

What does the term common rail refer to?

....

....

Identify the parts indicated.

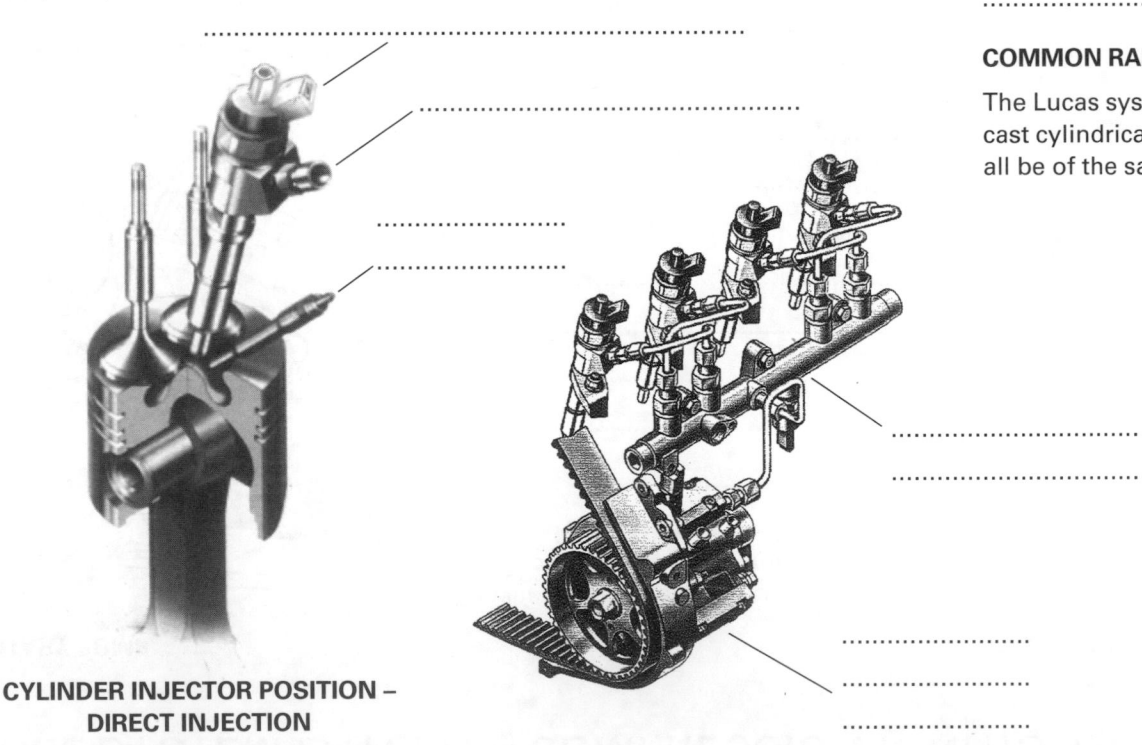

**CYLINDER INJECTOR POSITION –
DIRECT INJECTION**

The Bosch system (shown left), employs a high pressure pump which supplies near constant pressure to a large diameter manifold tube or 'rail'. This forms the pressure reservoir, feeding pipes of conventional diameter to the injectors. Electronic operation of the solenoid on each injector gives precise control over the timing and metering of fuel injected into each cylinder.

At what pressure does the system operate? ....

Where is the low pressure feed pump positioned? ....

How is direct injection diesel knock reduced by this system? ....

....

....

Name a common way to reduce the level of engine noise (i.e. diesel knock) heard by vehicle occupants.

....

COMMON RAIL FUEL INJECTION WITHOUT THE RAIL

The Lucas system (shown below) has no long tube or 'rail' but instead uses a cast cylindrical pot to which the injector pipes connect. NOTE: These pipes must all be of the same length.

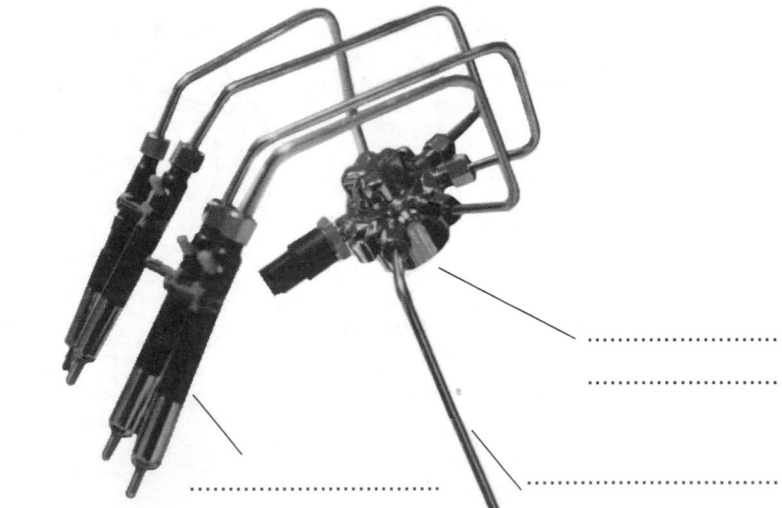

EXHAUST SYSTEM

The purpose of the exhaust system is to silence the noise created by the high velocity of the exhaust gas as it leaves the engine.

To remove harmful engine pollutants modern exhaust systems use a catalytic converter. When operating the catalytic converter acts as a 'red hot furnace' and causes the harmful gases to reburn, react with one another and render themselves harmless.

Carbon Monoxide burns to ....

Hydrocarbons (unburnt fuel) burns to ....

Oxides of Nitrogen (NO$_x$) burns to ....

What precautions should be observed to ensure long catalytic converter life?

1. ....

2. ....

3. ....

4. ....

A complete exhaust system is shown

Name the parts indicated.

Name the items shown below

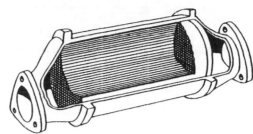

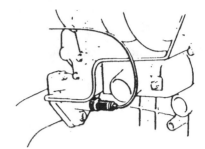

....

State the purpose of the lambda sensor?

....

....

....

Examine sectioned silencer boxes similar to the types shown and describe how they differ.

Absorption type

....

....

....

Expansion (or baffle) type

....

....

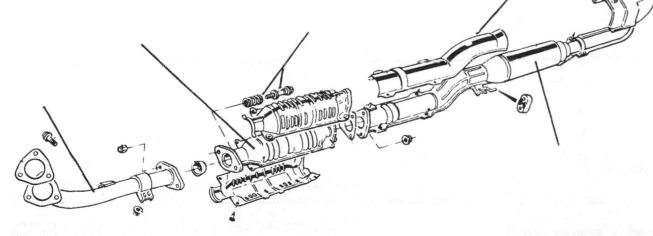

CLUTCH

The clutch is a form of coupling which is used to connect the engine crankshaft and flywheel assembly to the gearbox input or primary shaft.

One main function of the clutch is to allow the drive to be taken up gradually and smoothly as the vehicle moves off from rest.

State two more functions of the clutch:

1. ....

....

2. ....

....

Single-plate Clutch

There are many different types of clutch in use on road vehicles but by far the most popular is the single-plate friction type. This type of clutch is operated by the driver depressing and releasing a pedal; it is used in conjunction with a manual type gearbox.

When a vehicle moves off from rest the clutch must make a connection between the rotating engine crankshaft and the stationary gearbox primary shaft.

Describe how clutch engagement allows the vehicle to move off smoothly.

....

....

....

....

....

Name the type of clutch used with an automatic gearbox.

....

Main features of operation (single-plate clutch)

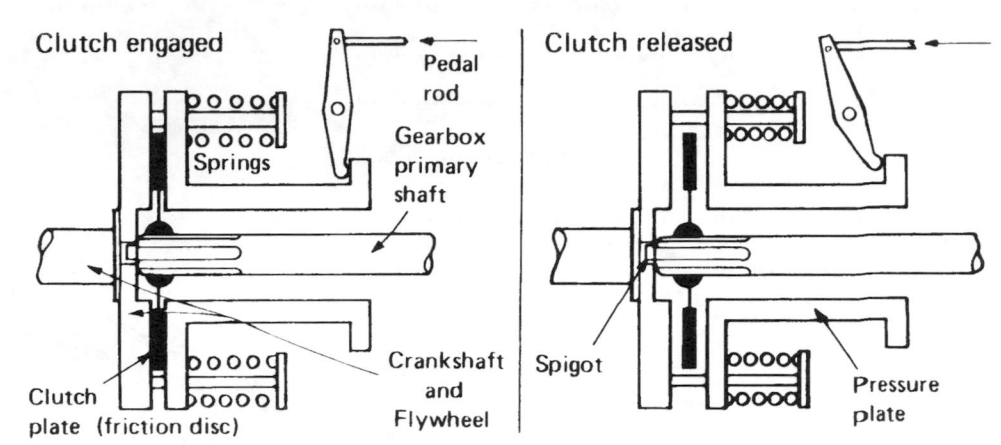

A very much simplified arrangement of a single-plate clutch is illustrated above.

Study the two drawings and state:

1. How the drive is transmitted from the engine to the gearbox.

....

....

....

....

....

2. How the drive to the gearbox is disconnected when the clutch pedal is depressed.

....

....

....

State the most important single factor with regard to the transmission of drive through this type of clutch:

....

The Diaphragm Spring Clutch

As an alternative to using a number of coil springs to provide the clamping force, many clutches use a single diaphragm type spring. The diaphragm spring is rather like a saucer in shape with a hole in the centre.

A single-plate clutch incorporating a diaphragm spring is shown below; complete the labelling on the drawing.

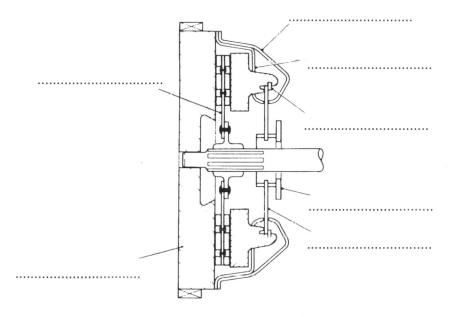

In this type of clutch the diaphragm spring serves two purposes:

(1) *It provides the clamping force.*

(2) ....

Name the type of clutch assembly shown below and label the drawing:

....

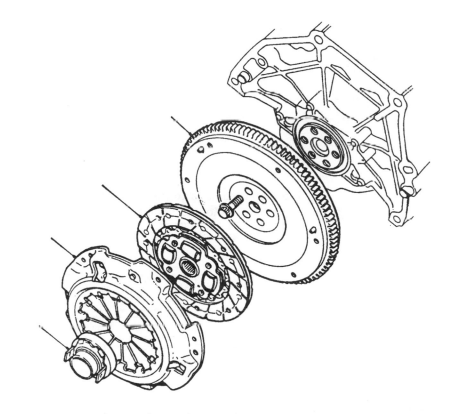

Why is it necessary to provide some form of adjustment (manual or automatic) in the clutch operating system from the clutch pedal?

....

GEARBOX – FUNCTION

In a motor vehicle the gearbox serves three purposes:

1. *To multiply (or increase) the torque (turning effort) being transmitted by the engine.*

2. ....

3. ....

Under many operating conditions the torque requirement at the driving wheels is far in excess of the torque available from the engine.

State four operating conditions under which the engine torque would need to be multiplied at the gearbox:

1. *When the vehicle is heavily laden.*

2. ....

3. ....

4. ....

Types of gearbox

The FOUR types of gearbox are:

1. *Sliding mesh* ....   2. *Constant mesh* ....

3. ....   4. ....

Which of the four types of gearbox are used in most modern cars?

....

Originally gearboxes were nearly all of the sliding mesh type but these are now mainly confined to heavy applications such as tractors and some heavy goods vehicles. One interesting exception to this, however, is the use of sliding mesh gears on certain racing cars.

Types of gearing

The spur gear, the helical gear and the double helical gear are all types of gears used in gearboxes.

By observation in the workshop, complete the sketches below to show the tooth arrangement for each type.

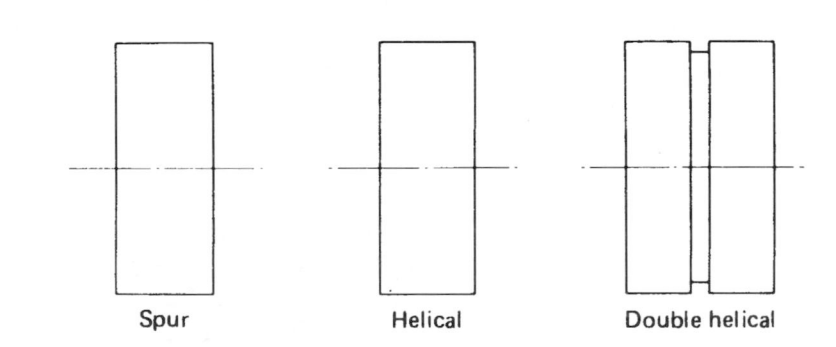

| Spur | Helical | Double helical |

Epicyclic gears

The sketch below shows a simple epicyclic gear train. This extremely compact arrangement of gears is used in:

.... and ....

Name the main parts of this gear train by completing the labelling on the drawing. Name the type of gearbox shown on the right.

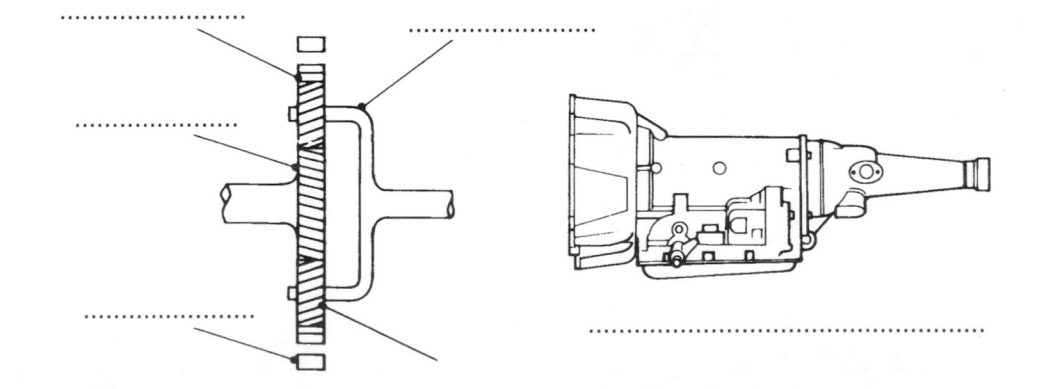

Simple Sliding-mesh Gearbox

The drawing below shows a three-speed sliding-mesh type of gearbox. To obtain the various gear ratios the gearbox layshaft is made up of different sized gearwheels which are connected in turn to the gearwheels on the mainshaft. The drive to the layshaft is via a pair of gearwheels which are permanently meshed.

Complete the labelling on the drawing:

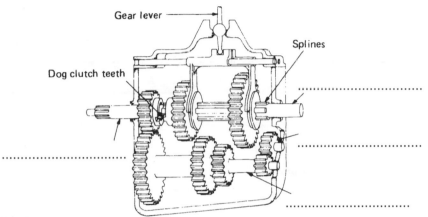

Gear lever

Splines

Dog clutch teeth

Investigation

Examine a sliding-mesh gearbox and answer the following questions:

(a) What type of gearing is used? ....

(b) How are the gears engaged to provide the various gear ratios?

....

....

....

....

(c) Why is the mainshaft splined?

....

....

(d) In what position is the gear lever in the box shown above? ....

(e) Show the power flow in 1st, 2nd and top gear by adding the mainshaft gearwheels and arrows to the drawings below. On the wheels in (4) show the direction of rotation; view the gear train from the rear of the gearbox.

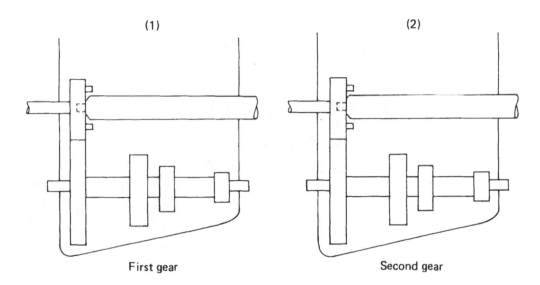

(1) First gear

(2) Second gear

(3) Top gear

(4) Reverse

Mainshaft

Idler gear

Layshaft

TRANSVERSE ENGINE – FRONT WHEEL DRIVE TRANSMISSION LAYOUT

This is a typical front wheel drive layout for a small car. State three vehicles which use this (or very similar) layout and name the numbered transmission parts.

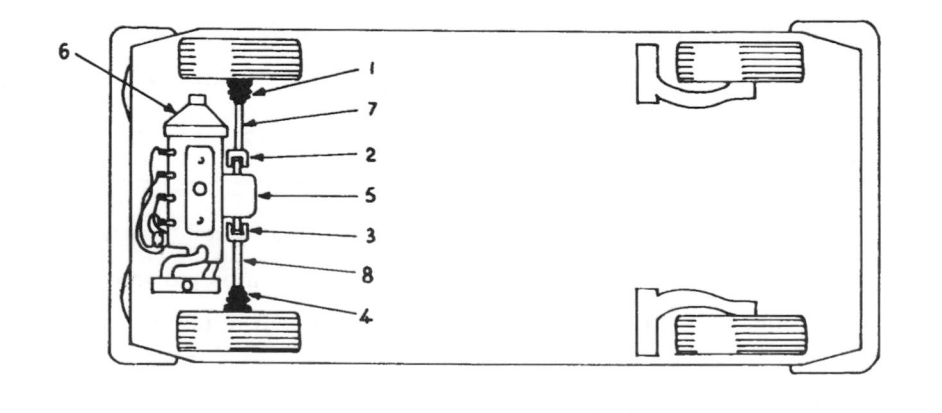

Make	Model	Engine size

1. ....

2. ....

3. ....

4. ....

5. ....

6. ....

7. ....

8. ....

COMBINED GEARBOX AND FINAL DRIVE

The drawing shows a gearbox, differential and inner constant velocity joints; these joints are covered by their rubber shrouds.

Name the main parts and show with arrows the power flow from the first motion shaft to the drive shafts, assuming top (4th) gear to be selected.

PROPELLER SHAFTS AND DRIVE SHAFTS

On a vehicle of conventional layout the purpose of the propeller shaft is to

transmit the drive from the.................................... to the ....

.... Drive shafts (or half shafts) transmit the drive from

the .... to the ....

(a) Give three examples of vehicle layouts in which external drive shafts may be used as opposed to conventional type half shafts enclosed in the axle casing.

1. *Front engine, rear wheel drive with 'independent' rear suspension.*
....

2. ....

3. ....

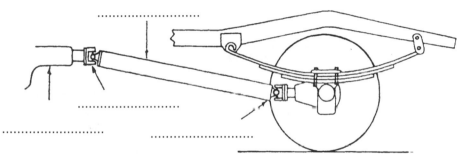

(b) Complete the labelling on the 'open' type propeller shaft arrangement shown above.

(c) Why is it necessary to have a universal joint at both ends of the propeller shaft?

....

....

....

Universal joints are also used on drive shafts to allow for the rise and fall of the road wheels relative to the final drive assembly on cars with independent rear suspension.

As the rear axle swings up and down with spring deflection the distance between the gearbox and axle varies. It is therefore necessary to enable the propeller shaft effectively to vary in length. The splined '*sliding joint*' in a propeller shaft assembly provides this facility.

Investigation

Examine a vehicle and make a sketch below to show the propeller shaft sliding joint arrangement.

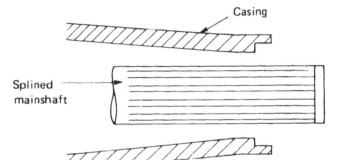

(a) Is the propeller shaft solid or tubular?

....

(b) State three advantages to be gained by using a tubular shaft.

1. ....

....

....

2. ....

3. ....

The Final Drive

The final drive consists basically of a pair of gears which provide a permanent gear reduction, thereby multiplying the torque being transmitted from the gearbox.

In most conventional transmission arrangements (for example front engine with rear live axle) the final drive fulfils another purpose; this is

....

....

The two gears forming the final drive are named.

1. .... 2. ....

Investigation

Examine a rear axle and complete the drawings below to show how the final drive gears are positioned relative to each other to give the desired motion.

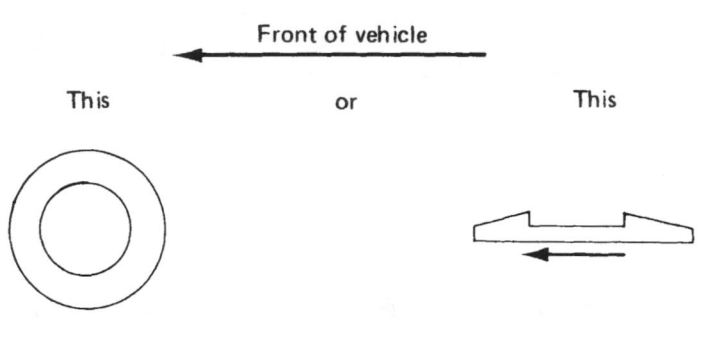

Complete the table below for the axle you examine.

Number of teeth on crown wheel	Number of teeth on pinion	Final drive gear ratio

Final Drive Bevel Gears

Early motor vehicles used the bevel gears shown at **1**.

1

These were noisy and relatively weak for motor vehicle work.

....

As you can see, the gears at **2** have different shaped gear teeth and are quieter in operation and stronger than type **1**.

2

Name these bevel gears.

....

Modern vehicles use the bevel gear shown below. Name the type of gear shown and state why it is now used in preference to the gears at **1** and **2**.

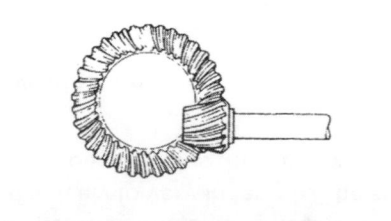

....

....

....

....

....

The Differential

When a vehicle is cornering the inner driven wheel is rotating more slowly than the outer driven wheel. It is the *differential* which allows this difference in speed to take place while at the same time transmitting an equal driving torque to the road wheels,

Why does this difference in speed between inner and outer road wheels need to occur?

....

....

....

The differential is normally an assembly of bevel gears housed in a casing (or cage) which is attached to the crown wheel. The essential parts of a differential are:

1. *Planet wheels* ....   3. ....

2. ....   4. ....

The differential could be described as a *torque equaliser*. Why is this?

....

....

....

During cornering the inner driven wheel is rotating at 100 rev/min and the outer driven wheel is rotating at 200 rev/min. The torque in the half shafts would be

(a) double in the outer

(b) the same in each

(c) double in the inner

Answer (   )

Investigation

Examine a final drive and differential assembly and complete the drawing below by adding the main components of the differential; clearly label each component.

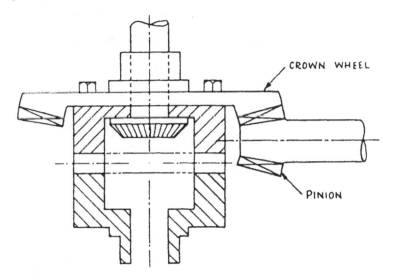

Which gearwheels transmit the drive to the axle shafts?

....

Power is transmitted from the final drive pinion to the axle shafts via the components listed below. Rearrange the list in the correct sequence in accordance with the power flow through the differential.

Components	Power flow
Cross pin	1. ....
Sunwheels	2. ....
Crown wheel	3. ....
Planet wheels	4. ....
Axle shafts	5. ....
Pinion	6. ....
Differential cage	7. ....

WHEELS AND TYRES

The road wheel assembly (tyre and metal wheel) transmits the drive from the drive shafts to the road surface, and back the opposite way during braking. It also provides a degree of springing to accommodate minor road irregularities.

Types of Road Wheel

Various designs of road wheels are in use on both cars and goods vehicles. On cars the pressed steel type of wheel has been used more than any other.

STEEL WHEEL

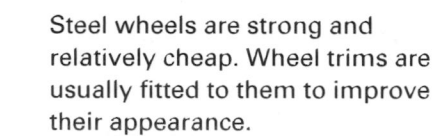

Steel wheels are strong and relatively cheap. Wheel trims are usually fitted to them to improve their appearance.

Name the two types of car wheel shown below and give reasons for using them.

....

   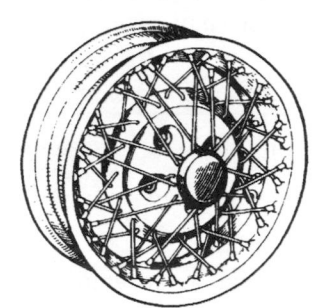

....

....

....

Road Wheel Attachment

The drawing below shows how the wheel nuts secure the wheel to the hub. You can see how the taper on the wheel nuts matches the taper provided around the stud holes in the wheel. What is the purpose of this taper?

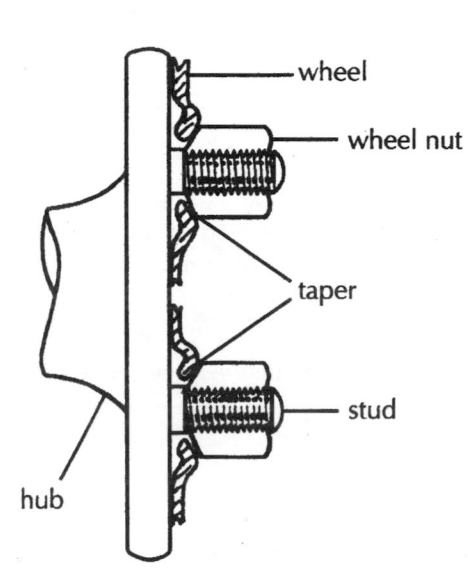

....

....

....

....

....

....

It is essential therefore that the taper on the nuts is fitted towards the wheel, otherwise the wheel would not be properly located on the hub.

An alternative wheel fixing is used on many modern cars, what is this?

....

State one more important thing to be aware of when fitting road wheels.

....

Tyres

The pneumatic tyre was originally developed for use on vehicles by J.B. Dunlop in the year 1888. Its main advantages over the solid tyres, in use at that time, were that it provided a cushion for the vehicle against road shocks and it rolled with greater ease.

However, pneumatic tyres also fulfil a number of other functions:

1. ....
2. ....
3. ....
4. ....
5. ....

Tyre construction

The tyre is a flexible rubber casing which is reinforced or supported by other materials, for example, rayon, cotton, nylon, steel.

What makes a tyre suitable for its application?

....

....

....

....

Label the main parts on the tyre section shown below:

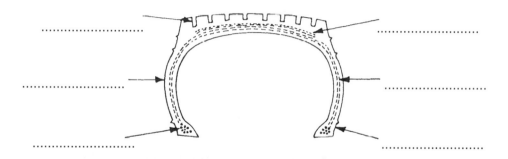

Tubed and Tubeless Tyres

In a tubed tyre and wheel assembly, the inner tube, together with its valve provides an air tight seal. This section through a tubeless tyre and rim shows how the tyre, rim and valve combine to form an air tight seal.

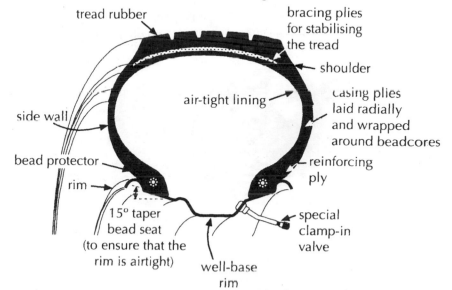

The drawings below show tubed and tubeless tyre valves. Label the parts of the sectioned valve.

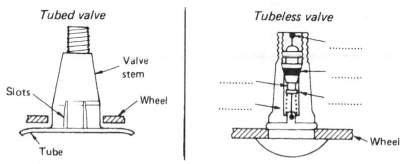

Tubed valve   *Tubeless valve*

What is the purpose of the slots in the outer stem of the tubed valve?

....

....

State three advantages of tubeless tyres over tubed tyres:

....

....

....

....

....

....

....

Types of tyre construction

The two principal types of tyre construction used on road vehicles are:

1. ....

2. ....

A tyre casing consists of *plies* which are layers of material looped around the beads to form a case. The basic difference in structure between radial and diagonal (or cross) ply tyres is in the arrangement of the casing plies.

With diagonal (or cross) ply construction the ply 'cords' form an angle of

approximately .... to the tyre bead

Radial ply cords form an angle of

....to the tyre bead.

Radial ply tyres are in use on most cars and on many goods vehicles.

Add lines to the two outlines at the top opposite to show the position of the cords in relation to the tyre bead.

Radial ply   *Diagonal (or cross) ply*

Label the tyre sections illustrated below.

....

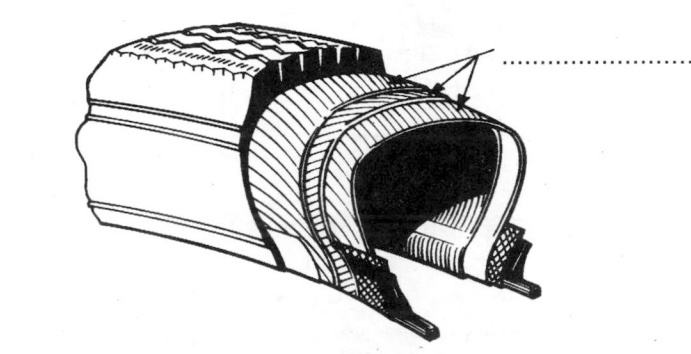

....

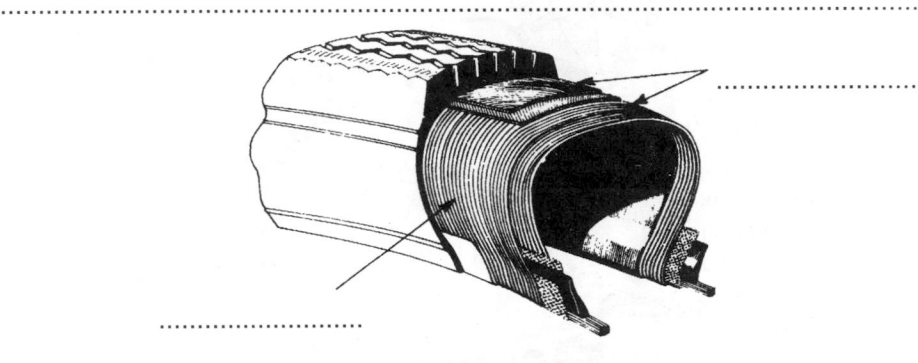

....

One feature of the radial ply tyre is that the walls are more flexible than the diagonal (or cross) ply tyre. Why is this?

....

....

....

Tyre Faults

Certain tyre defects make it illegal to use a vehicle on a public road. Outline the main legal requirements with regard to the tyre faults listed below.

Tread wear

....

....

....

....

Cuts

....

....

....

....

Lumps and bulges

....

....

....

....

Investigation

Examine the tyres on a vehicle and complete the table below.

Tyre Size Markings

The size of a tyre is indicated within markings on the sidewall, for example 185/60 R14 82H.

Complete the table below to show what is meant by each of these markings and indicate on the drawing the positions of the two main dimensions.

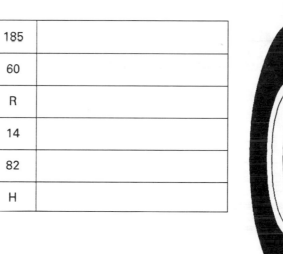

185	
60	
R	
14	
82	
H	

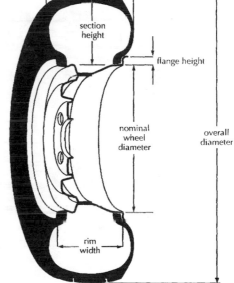

Position	Tread depth		Pressure	Other defects	Serviceability
	Maximum	Minimum			
Front O/S					
Rear O/S					
Front N/S					
Rear N/S					
Spare					

SUSPENSION SYSTEMS

The suspension on a vehicle performs two functions:

1. It insulates the vehicle body, hence the passengers or load, from shocks as the vehicle travels over irregularities in the road surface.

2. It assists in keeping the tyres in close contact with the road surface to ensure adequate adhesion for accelerating, braking and cornering.

One of the simplest and most widely used forms of suspension was the *beam axle semi-elliptic leaf spring* arrangement. This system is still used at the rear of some older cars and at both the front and rear of most heavy commercial vehicles.

Beam Axle Semi-elliptic Leaf Spring Suspension

(a) Complete the labelling on the drawing:

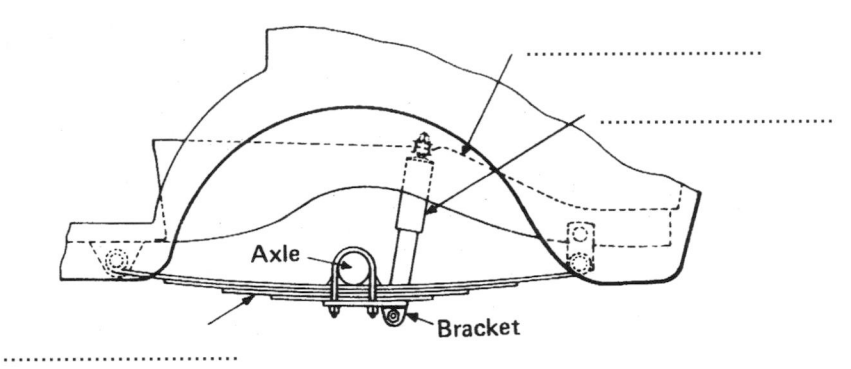

Briefly describe the action of this suspension system.

....

....

....

....

....

(b) Name and briefly describe the four suspension conditions illustrated below:

....

....

....

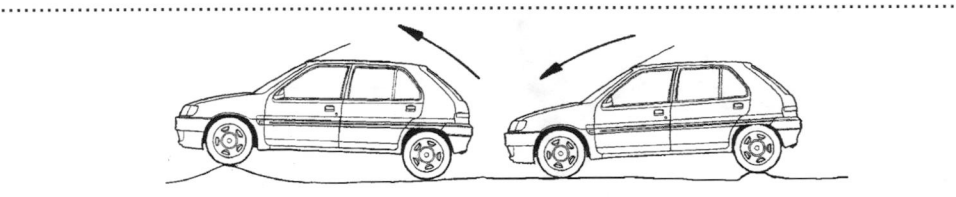

....

94

The Laminated Leaf Spring

The drawing below shows a leaf spring. Label the drawing and state the purpose of each part.

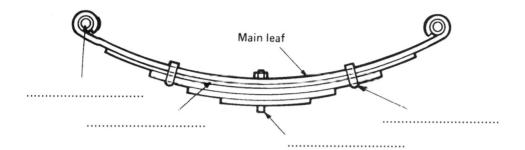

Main leaf

....

....

....

Leaves

....

....

Springclips

....

Centre bolt

....

....

Spring eye bushes

....

Why is this type of spring said to be semi-elliptical?

....

....

....

The simplified sketch (a) below shows how the leaf spring is attached to the chassis.

State the purpose of the swinging shackle and complete sketch (b) to show the action of the shackle as the spring is deflected:

....

....

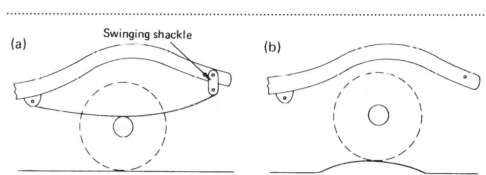

(a) Swinging shackle (b)

Investigation

(a) Examine a leaf spring and axle assembly (car type) and make a sketch to show how the spring is secured and located on the axle.

(b) Sketch and label the spring eye bush.

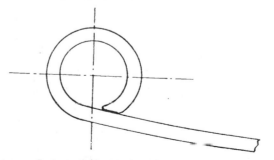

Helical Springs

State the advantages of a helical or coil spring over a laminated leaf spring.

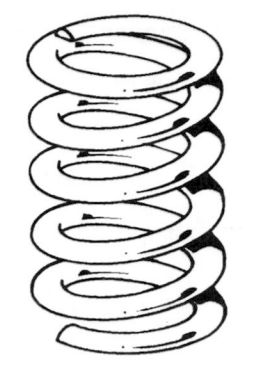

Advantages

1. ....

2. ....

3. ....

4. ....

5. ....

As the term implies, 'independent suspension' describes a suspension system in which the suspension at one wheel operates completely independently of the opposite wheel on the other side of the vehicle. This is not the case with a beam axle suspension when, if one wheel rises over a bump, the other wheel on the same axle is also affected.

Complete the sketch below to show how the other wheel is affected as one wheel rises over a bump with beam axle suspension.

View of axle from front

Independent Suspension

Vehicle Body

The drawing above is a simplified representation of a double-link independent suspension system. Complete the drawing to show the springs and the suspension as one wheel rises over a bump.

Name the suspension types shown below and label the drawings.

Give examples of current vehicles using these suspension types:

Make	Model	Make	Model
....	....	....	....

(a) State the meaning of the terms i.f.s. and i.r.s.

i.f.s. ....

i.r.s. ....

(b) State the advantages of independent suspension over beam-axle suspension.

1. ....

2. ....

3. ....

4. ....

5. ....

The main disadvantages of independent suspension are that it is usually more complicated and costly and the tyres tend to wear unevenly.

(c) In the suspension arrangement shown below 'S' is the spring. Complete the labelling on the drawing and name this type of spring.

....spring.

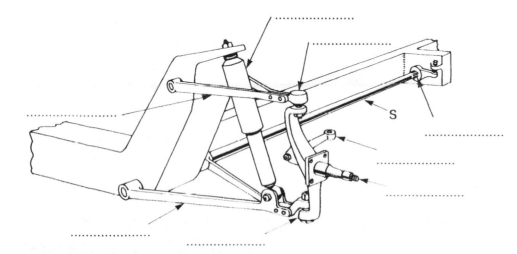

(d) Name the type of spring in each of the suspension arrangements shown below.

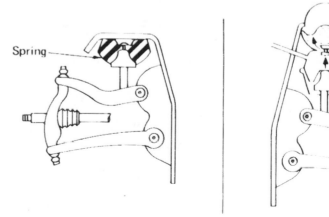

....

(e) The suspension shown with the pneumatic form of spring is a 'hydro-pneumatic' suspension. Why is it so called?

....

....

....

(f) In the hydro-pneumatic suspension system the individual suspension units are interconnected. How are they interconnected?

....

(g) Give examples of vehicles which use the types of suspension springs illustrated on this page.

Vehicle make	Model	Type of spring

97

Suspension Dampers

Why is it necessary to use dampers in a suspension system?

....

....

....

....

....

....

The dampers used on very many cars are the hydraulic telescopic type.
Telescopic dampers are fixed at one end to the axle or suspension link and
at the other to the

....

Principle of operation

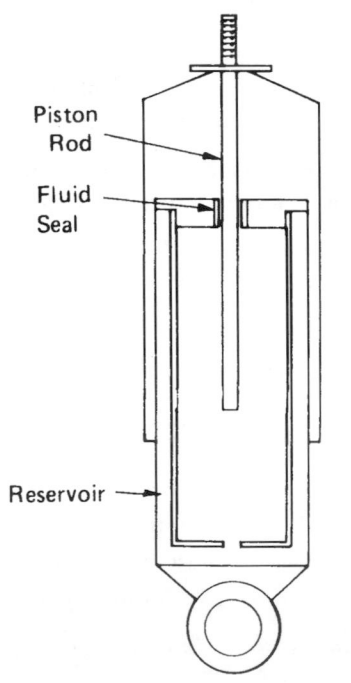

Piston
Rod

Fluid
Seal

Reservoir

Complete the simplified drawing of the
telescopic damper shown.

As the damper is operated fluid is forced from
one side of the piston to the other through small
drillings. The limited rate at which fluid can flow
creates a resistance to movement which
prevents undue flexing of the road spring.

Why does some fluid flow into and out of the
reservoir at the base of the damper?

....

....

....

....

GOODS VEHICLE SUSPENSION

As already mentioned, some goods vehicles use semi-elliptic suspension
springs. Some heavy goods and passenger vehicles use quite different systems.
for example, pneumatic (air) suspension.

Air Suspension

Air springs usually consist of reinforced rubber bellows containing air under
pressure, situated between the axle, and an air capacity tank mounted on to the
chassis. They have an important advantage in that by fitting levelling valves to
each wheel unit, the height of the vehicle can be kept constant whatever
the load.

The main features of air suspension are shown below; label the drawing.

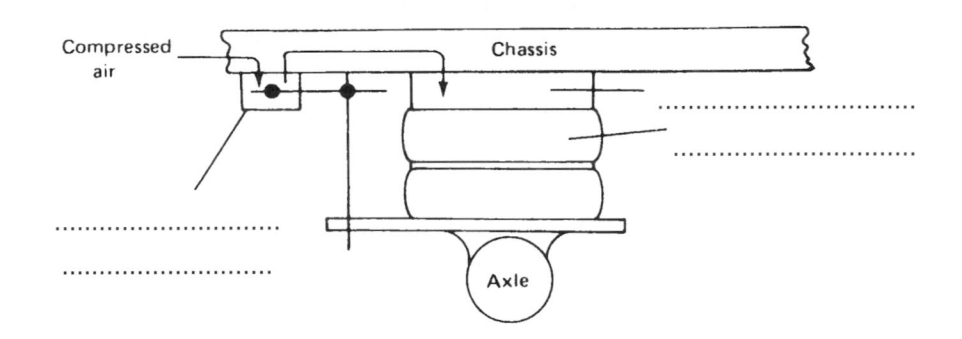

By observation, list three vehicles using air suspension.

Vehicle make	Model	Gross vehicle weight	Air suspension manufacturer

BEHAVIOUR OF MATERIALS UNDER APPLICATION OF FORCE

A force tends to produce or alter the motion of a body if it is free to move, or induce an internal reaction – called *stress* – in a body if it is not free to move.

When a bolt is tightened the applied force will always cause the material to deform – the material is thus said to be under *strain*. Excessive strain will cause permanent deformation.

Below are the main forces which act on a material.
State two motor-vehicle components subjected to each force.

Force	Motor-vehicle components
Tensile or stretching	
Compressive	
Torsional or twisting	
Shearing	

How would permanent deformation caused by overtightening affect nuts and bolts or studs?

....

....

....

....

Use arrows to indicate the directions in which the major forces would normally be acting after assembly when the vehicle is in normal use.

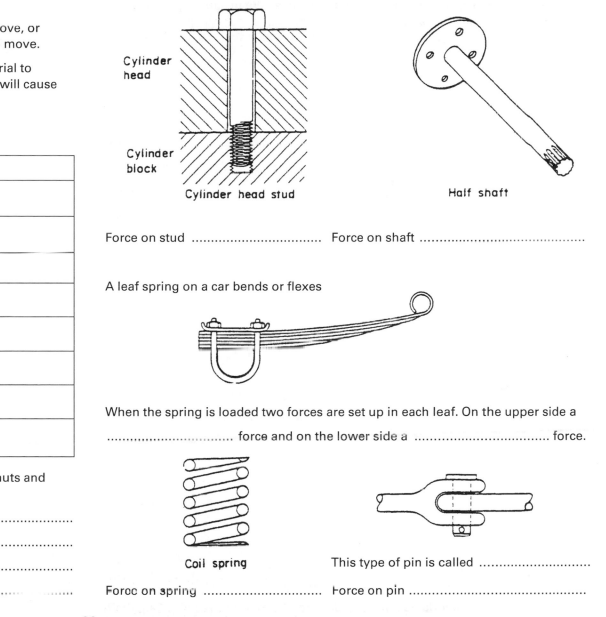

Cylinder head

Cylinder block

Cylinder head stud

Half shaft

Force on stud .... Force on shaft ....

A leaf spring on a car bends or flexes

When the spring is loaded two forces are set up in each leaf. On the upper side a .... force and on the lower side a .... force.

Coil spring

This type of pin is called ....

Force on spring .... Force on pin ....

99

STEERING SYSTEMS – TYPICAL BEAM AXLE LAYOUT

The front wheels on a normal vehicle rotate on stub axles which swivel or pivot about what is known as the 'swivel axis'. This provides a means of steering the vehicle. To make the stub axles pivot, a control gear, operated by the driver via the steering wheel, and a linkage to the stub axles are necessary.

One of the simplest steering arrangements is that used with a beam axle system. Complete the beam axle steering layout below and label each part.

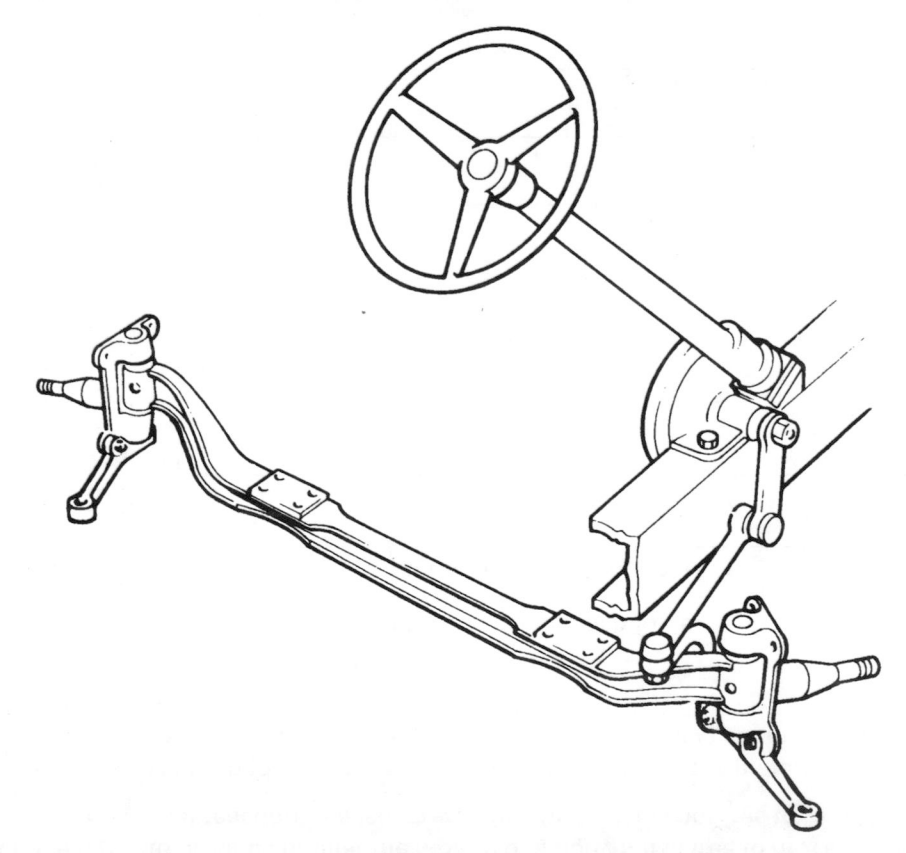

On what type of vehicle is the system shown above normally found?

....

State the purpose of each of the following steering components:

Steering gearbox

....

....

....

....

Drop arm

....

....

Draglink

....

....

Steering arms

....

....

Track rod

....

....

Ball joints

....

....

....

The *steering swivel axis* in the arrangement shown opposite is formed by 'king pins' and 'brushes'. That is, the stub axles pivot on hardened steel king pins which are held firmly in the beam axle; bushes in the stub axles provide the necessary bearing surfaces.

Car Steering System

Most cars have independent front suspension (i.f.s.), and the steering linkage must therefore be designed to accommodate the up and down movement of a steered wheel without affecting the other steered wheel. In most systems two short track rods, which pivot in a similar arc to the suspension links, are connected to the stub axles.

One system in use on cars can be described as a steering gearbox with 'idler' and three track rod layout.

Complete the drawing of such a system shown below by adding the linkage and labelling the parts.

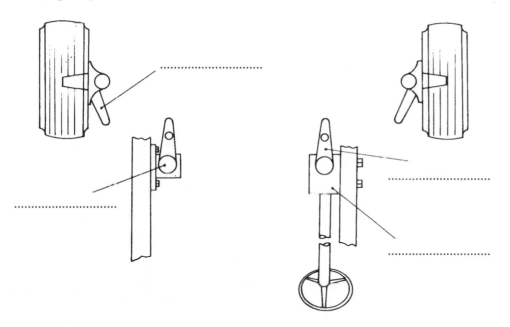

What are two main disadvantages of this layout?

....

....

Name the steering system shown below and label the parts.

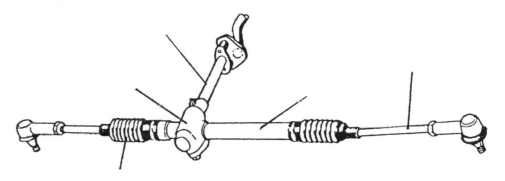

....

Draw a centre line through the steering swivel axis on the stub axle arrangement shown below.

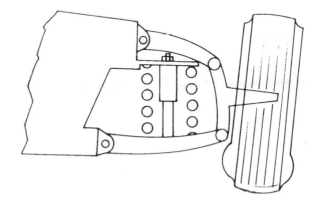

What provides the steering swivel action on this arrangement?

....

101

CENTRE OF GRAVITY, EQUILIBRIUM AND STABILITY

On the vehicle shown below the cross marks the *centre of gravity*. With a front engine, front wheel drive vehicle the C of G would be nearer to the front.

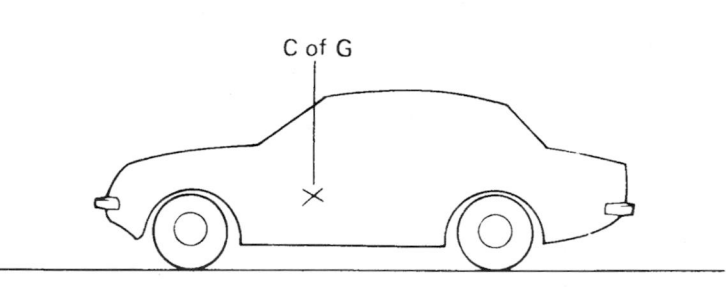

In the space below, sketch a vehicle that would have its C of G towards the rear, and indicate its approximate position.

The stability of a vehicle when cornering or braking is greatly affected by the actual position of the vehicle's C of G.

Give examples of vehicle types which have:

1. Low C of G ....

2. High C of G ....

Consider the vehicle shown below. During braking the force *F* due to the inertia of the vehicle acting at the C of G tends to tilt the vehicle about the front wheels, thus putting more load on to the front wheels, and reducing the load on the rear wheels 'Load transfer' takes place.

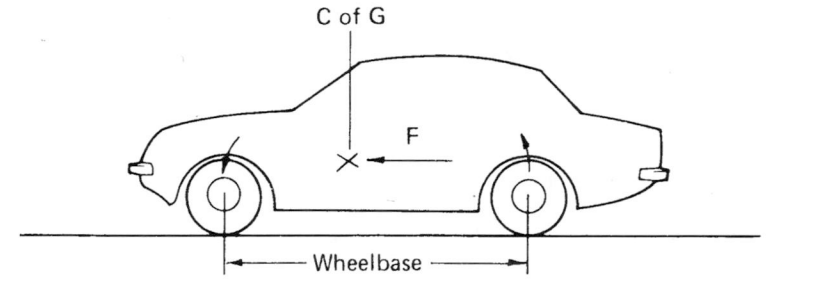

The degree of load transfer is dependent upon a number of factors. How do the following affect load transfer during braking?

1. Height of centre of gravity

   ....

   ....

2. Wheelbase

   ....

   ....

3. Rate of retardation

   ....

   ....

4. Ratio of C of G height to wheelbase

   ....

   ....

BRAKING SYSTEM

The braking system provides the driver of a vehicle with a means of safely slowing down (or retarding) the vehicle and bringing it to a halt, and preventing movement when stationary. When the

brakes are applied, ... is generated which converts the energy

of movement, ... energy, into .... energy.

Simple Mechanically Operated Drum Brake

The drawing on the left below shows a simple 'cam' operated drum brake in the 'off' position. Complete the drawing on the right to show the brake in the applied position.

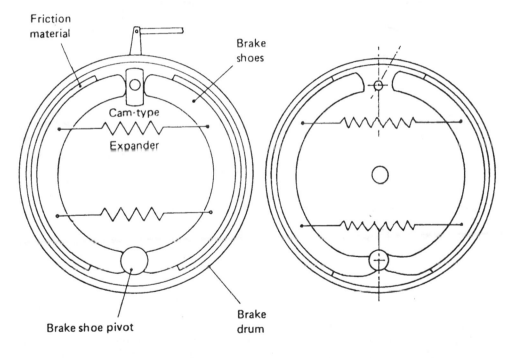

Which two methods are used to secure the friction material to the brake shoes?

....

Hydraulically Operated Brakes

Most modern car braking systems and many light commercial vehicle braking systems are operated hydraulically. Label the parts on the single-line hydraulic brake layout below.

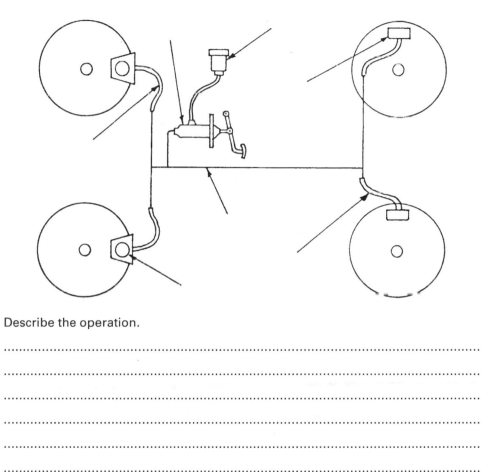

Describe the operation.

....

....

....

....

....

....

What is the reason for using flexible pipes in parts of the hydraulic system?

....

Single-leading Shoe Brake

The drawing shows a 'double-piston wheel cylinder' acting on a *leading brake shoe* and a *trailing brake shoe*. The leading shoe is pressed harder against the drum by drum rotation whereas the trailing shoe tends to be forced away by drum rotation. A much greater braking effect is therefore obtained from a leading shoe.

Name the parts in drum brake assembly.

Indicate drum rotation and show which is the leading and which is the trailing shoe.

Wheel Cylinder

Pull off springs

Brake shoe pivot

Why does the leading shoe wear at a greater rate than the trailing shoe?

....

....

....

What friction material may be used for brake linings?

....

What health hazards are associated with the dust from this friction material?

....

....

....

A typical rear drum assembly is shown below. State the purpose of the arrowed components.

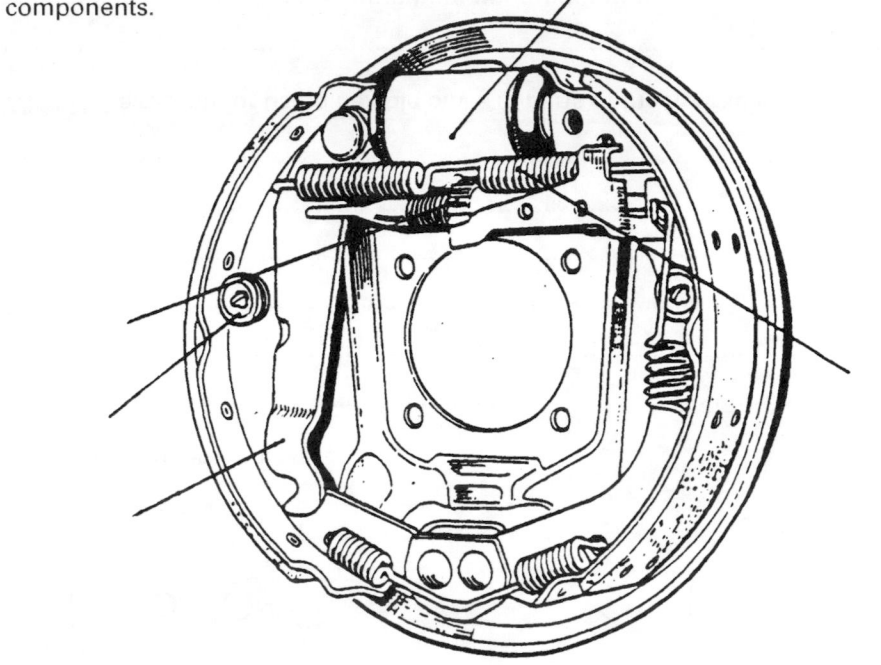

....

....

....

....

....

....

....

....

....

....

Disc Brakes

A disadvantage with drum brakes is that repeated brake applications at high speeds, for example, fast driving along winding roads or during long downhill descents, cause a gradual build-up in temperature of the brake assemblies particularly the linings and drums. Too great an increase in temperature reduces the efficiency of the brakes, making it more difficult to stop the vehicle. In effect the brakes may become temporarily useless.

This fall-off in brake performance is known as:

....

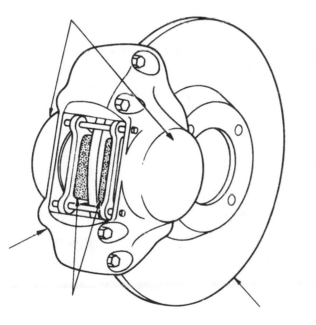

Investigation

Examine a front disc brake assembly and name the component to which the brake caliper is attached.

....

Disc brakes are now in use on most car front brakes and at both front and rear on many of the larger and faster cars. The disc brake, shown opposite, operates rather like a bicycle brake, that is friction pads are clamped on to a rotating disc by a caliper mechanism.

(a) Label the parts on the simplified disc brake shown opposite.

(b) State the purpose of the numbered components on the single-piston, sliding caliper shown below.

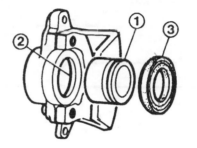

1. ....

....

2. ....

....

3. ....

....

A vacuum servo is part of the braking system on many modern cars, in particular vehicles with disc brakes. Why is the servo used?

....

....

....

....

....

Brake Compensation

A hydraulic braking system is said to be 'self-compensating'. What is meant by brake compensation and why are hydraulically operated brakes self-compensating?

....

....

....

....

....

....

....

....

Parking Brake (Handbrake)

A typical handbrake system is shown top right, operating drum or disc rear brakes. Label the drawing and state the purpose of component A.

....

....

....

....

Why is it necessary to provide some form of adjustment in a braking system?

....

....

....

....

Name the type of braking system shown diagramatically below:

....

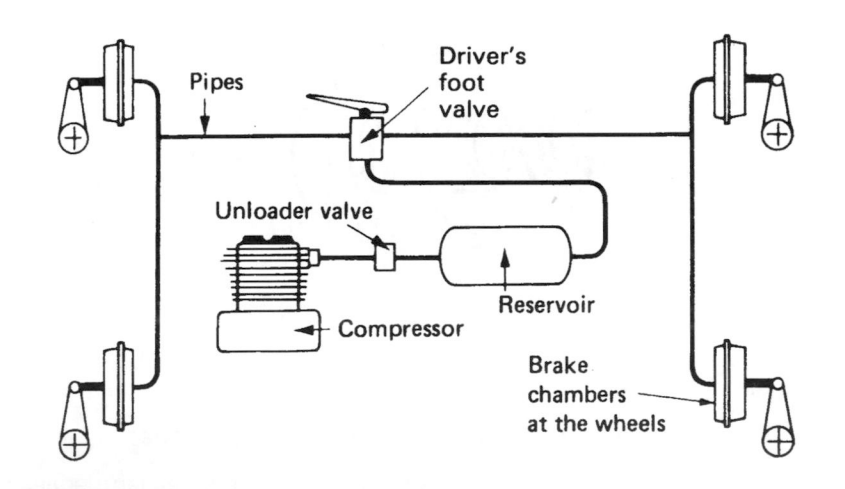

On which type of vehicle is this system normally found?

....

FRICTION

Friction is the resistance to motion, or reactive force, produced when two surfaces in contact are made to slide over each other.

Other effects of friction are:

(a) ....

(b) ....

The force F required to move the block over the surface shown below must be sufficient to overcome the frictional resistance. It is known as the *force of friction*. It is reactive force and the law of .... applies.

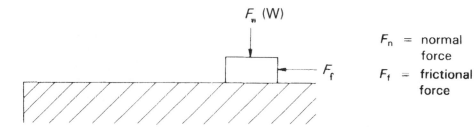

F_n = normal force

F_f = frictional force

(a) State the two factors which affect the force of friction:

(1) ....

(2) ....

Coefficient of friction

For any two surfaces in contact, the ratio F_f/F_n is known as the *coefficient of friction*, the symbol for which is the Greek letter μ (mu).

Hence $\mu = \dfrac{F_t}{F_n}$

The coefficient of friction is therefore an indication of the frictional 'quality' existing for any pair of surfaces in contact.

What single factor has the greatest effect on the coefficient of friction?

....

What is the effect on vehicle retardation caused by excessive use of the brakes? .

....

....

Problems

What force will be required to pull a spare parts packing case, which has a total downward load of 500 N, along a horizontal garage floor if the coefficient of friction between the case and floor is 0.5.

$$\mu = \frac{F}{W}$$

$$\therefore \qquad F = \mu \times W$$

1. If the parts, which have a mass of 400 N, are now removed, calculate the force required to pull the now empty packing case.

   ....

   ....

   ....

2. A vehicle exerts a downward load of 10 kN. If the coefficient of friction between the tyres and road is 0.55, the maximum retarding force which could be applied by the brakes would be

   (a) 5.5 kN   (c) 0.55 kN

   (b) 100 kN   (d) 55 kN

   Ans. (   )

3. The retarding force produced by the braking system of a vehicle when locking all four wheels is 6 kN. If the vehicle has a weight of 10 kN (1000 kg), calculate the coefficient of friction between the tyres and the road.

   ....

   ....

   ....

Chapter 4

Vehicle Electric/Electronic Systems

ELECTRICAL SYSTEMS

Electrical Current Flow

To allow an electrical current to flow an electric circuit must consist of

(a) A source of supply.

(b) A device that will use the supply to do useful work.

(c) Electrical conducting materials that will transfer the electric current from the supply source to the consuming device, and then return it to the supply source.

On a motor vehicle TWO sources of electrical supply are:

1. ....

2. ....

Name FIVE different types of devices that consume the current to do useful work:

1. ....

2. ....

3. ....

4. ....

5. ....

What is meant by the term 'electrical conductor'?

....

....

What is meant by the term 'electrical insulator'?

....

....

Name SIX electrical conductors and SIX insulators.

Conductors	Insulators
1.	1. ....
2.	2. ....
3.	3. ....
4.	4. ....
5.	5. ....
6.	6. ....

Earth Return

On conventional vehicles it is common practice to allow the current, once it has passed through the electrical resistance that it has operated, to return to the battery through the body frame, instead of by a separate cable.

The body frame therefore: ....

....

Sketch a wiring diagram showing four light bulbs connected in parallel, via a switch, to a 12 V battery. Incorporate the earth return symbol.

The advantages of the earth return system are:

1. ....

2. ....

3. ....

4. ....

Two types of vehicle that do not use the earth return system are:

1. ....

2. ....

They use a system called an insulated return.

....

....

The system is used because ....

....

ELECTRICAL SIGNS AND SYMBOLS

Each component on the list below can be represented by one of the symbols shown on this page. Write the appropriate name beneath each symbol.

Voltmeter

Switches

Battery

Ammeter

Light bulbs

Fuel pump

Winding

Ignition coil

Radio

Speaker

Starter motor with solenoid switch

Crossed wires not connected

Earth ground

Resistor fixed

Plug and socket connector

Multi plug and socket

Windscreen wiper motor

Contact-breaker points

Variable resistor

Heated rear window

Distributor cap and sparking plug

Crossed wires connected

Twin filament lamp

Gauge

Fuse

Diode

Horn

Alternator

Capacitor

Transistor

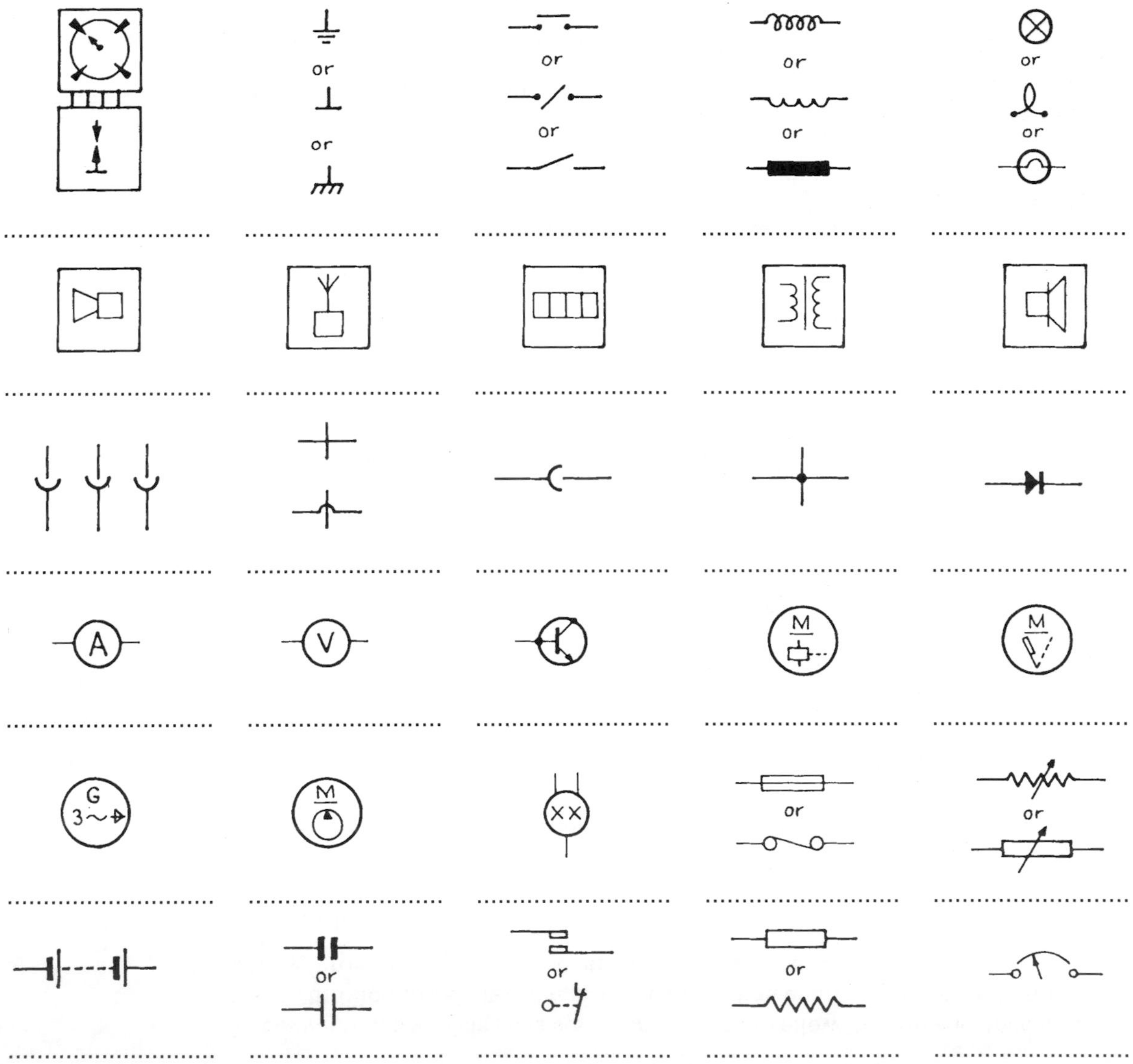

CABLES

The selection of cable and type of insulation required for motor vehicle use depends upon two main factors. These are:

1. ....
2. ....

Cables are often bunched together in a harness or loom. This simplifies fitting and ensures less chance of breakage or short circuits. Each cable within the harness is colour coded. What is the reason for colour coding?

....

....

The main feed cable to a specific switch may have a single colour and past the switch the lead has the same main colour but with a different coloured trace line passing through it.

Typical main colours for circuits are:

Ignition .... Headlamp ....

Sidelamp .... Flasher ....

The amount of current a cable can pass is determined by its size. How are these classified? ....

What does a cable having a size of 28/0.30 indicate?

....

What is a recognised safe current capacity for each strand of cable 0.30 mm thick? ....

Investigation

Examine cables of various thicknesses obtained from or still held in a wiring harness.
From what material are the wires made? ....

Count the number of strands and measure their diameter.

STARTER MOTOR

Starter motors fitted to cars are usually one of the two types shown. What is the purpose of the starter motor and what type of energy occurs?

....

....

Examine starter motors of the types shown and identify the arrowed parts.

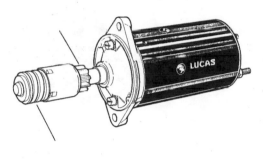 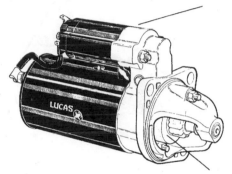

Type (I)

....

Type (II)

....

Explain the mechanical operation of Type (II) which is also shown opposite.

....

....

....

....

....

....

....

....

....

Name the parts indicated.

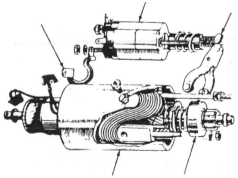 

LEAD-ACID BATTERY

The lead-acid battery is of the secondary cell type and is used on most automobiles in either 12 or 24 V form. The 12 V is the most popular. The battery acts as a reservoir of electrical power. All other components take current from the battery; that is, except the alternator (or dynamo) which charges the battery while the engine is running.

Construction

A 12 V battery container consists of six separate compartments. Each compartment contains a set of positive and negative plates; each set is fixed to a bar which rises to form the positive or negative terminal. The plates have a lattice-type framework into which is pressed the chemically active material.

Between each plate is an insulating separator.

Name the parts on the sketch below.

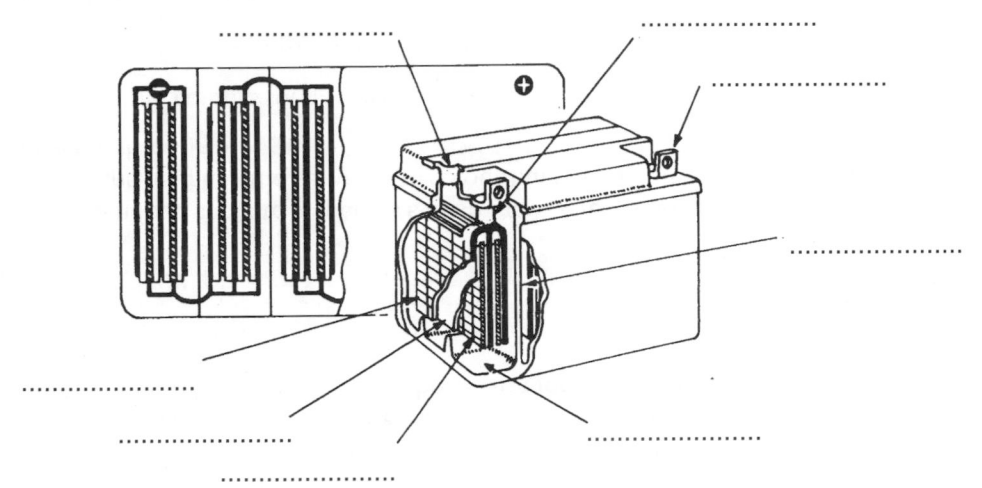

Investigation

Examine batteries suitable for dismantling. Dismantle a cell and state:

The colour of the active material.	The number of plates in each cell.
(a) Positive plate	(a) Positive ....
(b) Negative plate	(b) Negative ....

A battery is a potentially dangerous component. It contains dilute sulphuric acid, gives off gases while being charged, is heavy and is capable of supplying a high electrical current.

What safety precautions should be observed when handling batteries? (See COSHH safety precautions, page 11.)

1. ....

....

2. ....

....

3. ....

....

GENERATORS

The basic operation of a generator is to

....

Name the generators shown below.

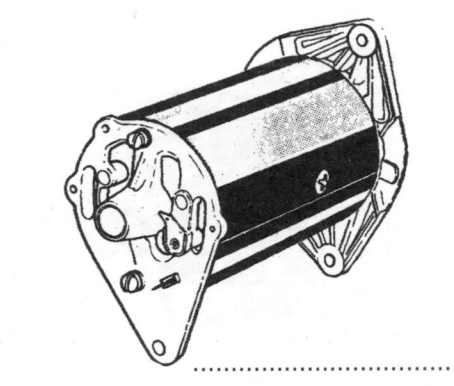

....

What advantages does the alternator have when compared with the dynamo?

....

....

....

IGNITION SYSTEM

The contact breaker point ignition system was used on vehicles for over sixty years and is commonly found on Classic Cars. Modern ignition systems are either electronic or computer controlled.

Name the main parts of the simple coil-ignition system shown and identify the primary and secondary circuits leads.

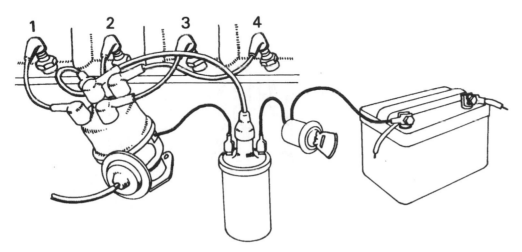

State the function of the following main components of a coil-ignition system:

Battery ....

....

Ignition switch ....

....

Coil ....

....

Distributor ....

....

Sparking plugs ....

....

The coil-ignition system must provide, at a precise time, a high voltage spark that will jump the plug gap inside the combustion chamber and ignite the compressed mixture.

The system may be considered to be made up of two electrical circuits:

1. Low-tension (primary) circuit – LT

2. High-tension (secondary) circuit – HT

The primary winding of the ignition coil is supplied at battery voltage (12 V) with a current of about 3 A. This supply builds up a magnetic field around both the primary and secondary coil windings.

Explain what happens to the output when the distributor contact breaker points are opened, or the electronic system breaks the circuit.

....

....

....

....

NOTE! Secondary voltage rises up to 12 000 V (or more) whilst the secondary current is about 0.005 A.

Complete the diagram and identify the basic symbols shown.

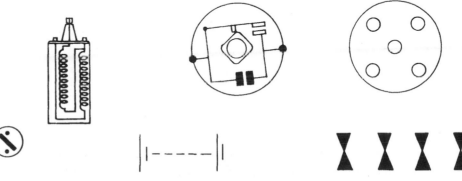

What is the firing order of the system shown, if the rotor turns anticlockwise?

....

SPARK PLUG AND IGNITION LEADS

Name the parts of
the spark plug
shown below.

Indicate the plug
thread diameter
and plug electrode
gap setting

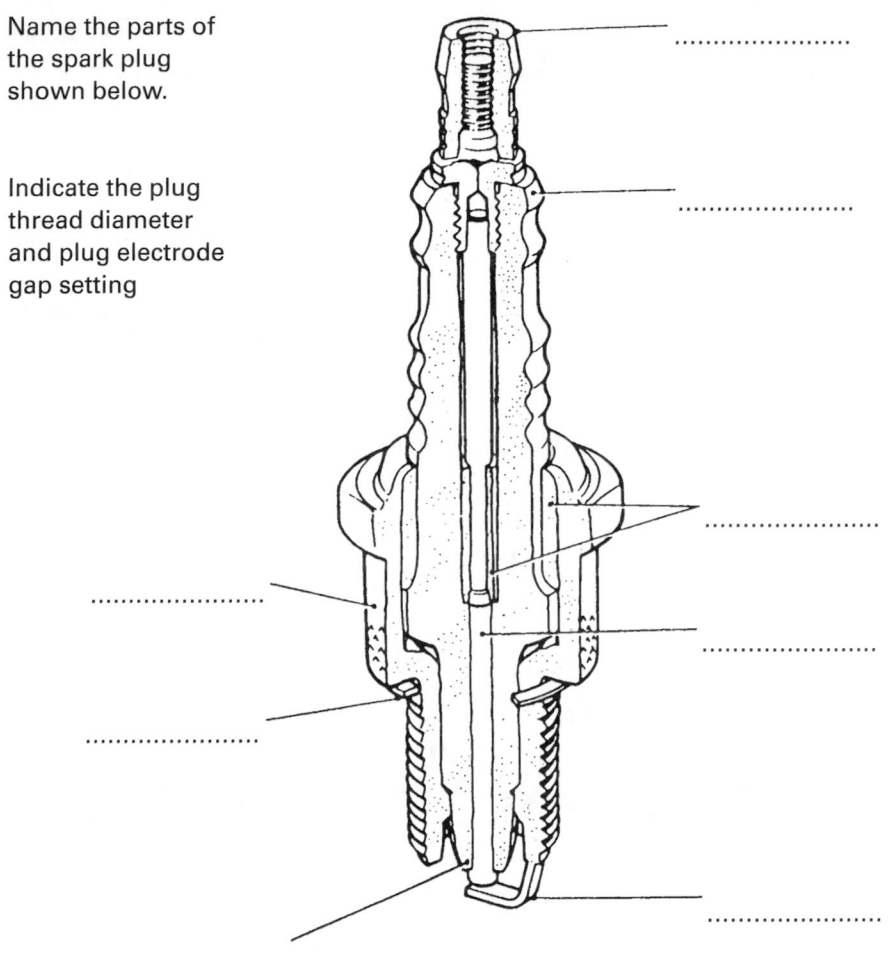

List the routine maintenance that should be given to a spark plug:

1. ....
2. ....
3. ....
4. ....

The spark plug consists of a centre metal electrode which passes through a form of ceramic insulator. The lower part of the insulator is fixed to a metal case which screws into the cylinder head. This case forms the earth to the engine.

At normal atmospheric pressure the plug will readily spark with little voltage, but when in the running engine the voltage requirement may progressively increase, for example, over a period it can rise from 7000 V to 15 000 V. State four factors affecting the voltage requirement of the plug:

1. ....
2. ....
3. ....
4. ....

What may cause a spark plug to foul and not spark?

1. ....
2. ....
3. ....
4. ....

Examine spark plug leads.
Two types of lead are commonly used. In one type the centre core is made of stranded wire.
What is used as the core of the second type of lead?

....

....

Both types are highly insulated. Why is this? ....

....

What electrical safety precaution should be observed when carrying out adjustments to the ignition system while the engine is running?

....

....

....

VEHICLE LIGHTING SYSTEMS

The lighting system of a car may be split into two basic circuits. These are:

....

List the basic essentials that make up a lighting circuit:

....

....

....

When more than one lamp is used the circuit is usually wired in ....

The side light circuit consists of two lamps at the front and at least two at the rear plus a lamp to illuminate the vehicle's rear registration plate. State their legal requirements.

....

....

....

The headlight circuit is two circuits, the main beam and dip beam. It also has:

....

....

....

Examine three vehicles of widely varying types, for example, commercial vehicle, Land Rover, and a modern saloon. Observe the correct working of head, dip and sidelights. If lights operate correctly, place tick in column. If inoperative, state which light is faulty.

Vehicle make			
Model			
Sidelights			
Tail lights			
Number plate light			
Headlight main beam			
Headlight dip beam			

Complete the following diagrams using the earth return system.

Sidelight circuit

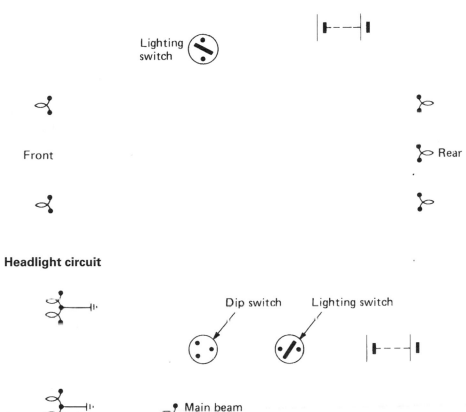

Headlight circuit

Other circuits, which are sometimes considered as part of the basic lighting system, but which use lights for signalling purposes or for better visibility are the:

1. ....

2. ....

3. ....

TYPES OF BULB

Examine light bulbs of different types.
Identify those shown and state where these may be fitted on a vehicle; in each case give a typical wattage rating.

Note: These sketches are not to scale.

1. ....

2. ....

3. ....

4. ....

5. ....

6. ....

7. ....

The inert gas used inside the bulb is usually ....

The resistance wire used for the filament is ....

The temperature reached by this wire is approximately ....

In what important way does bulb 3 differ from the other small bulbs?

....

....

....

Why does bulb 4 have a staggered pin fitment? ....

....

....

Bulbs 5 and 6 are known are pre-focus bulbs. What identifying feature gives them this name?

....

....

....

....

What does unit 7 include? ....

....

For the same wattage rating the headlamp bulb 5 will give off a much brighter light than bulb 6. How is this improvement achieved?

....

....

....

....

Methods of Dipping the Beam

Two-headlamp system

When on main beam the light rays illuminate the road as far ahead as possible.

What occurs when the lights are dipped?

....

....

....

Show the position of the dipped beams.

Four-headlamp system

With some systems, when on main beam the outer lamps are on permanent dip while the inner lamps throw out a brighter longer range beam along the road.

What occurs when the lights are dipped?

....

....

Show the position of the light beams when on main beam.

The direction indicator, stop light, reversing light and rear fog light are all obligatory lighting/warning systems on a modern car. Complete the simple wiring diagrams of the circuits shown.

Direction Indicator

Include flasher unit panel indicator lamps and, if required, side indicators.

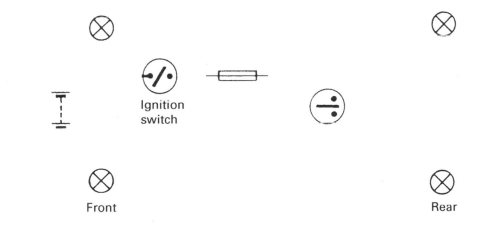

Front   Rear

Stop Lamps

Indicate how these are operated.

Ignition
switch

117

SYMBOLS FOR CONTROLS, INDICATORS AND TELL-TALES

Modern vehicles now use international standard symbols on control knobs, switches, fuse units, etc. to indicate what each item does. The symbols are designed to give the driver a clear pictorial indication of the function of each item. The number of these symbols seems to be inexhaustible and it can sometimes be difficult to interpret what they all indicate. Examine the examples on this page, and use information from other sources to familiarise yourself with these types of symbols.

NOTE: If the symbol is in side view (i.e. lamps) assume the vehicle is being driven from right to left; and if a plan view is shown, the vehicle is being driven 'upwards' or forwards.

A fuse/relay box from a FORD vehicle is shown below. Complete the tables opposite to state what the relays operate and what main circuits the fuses protect.

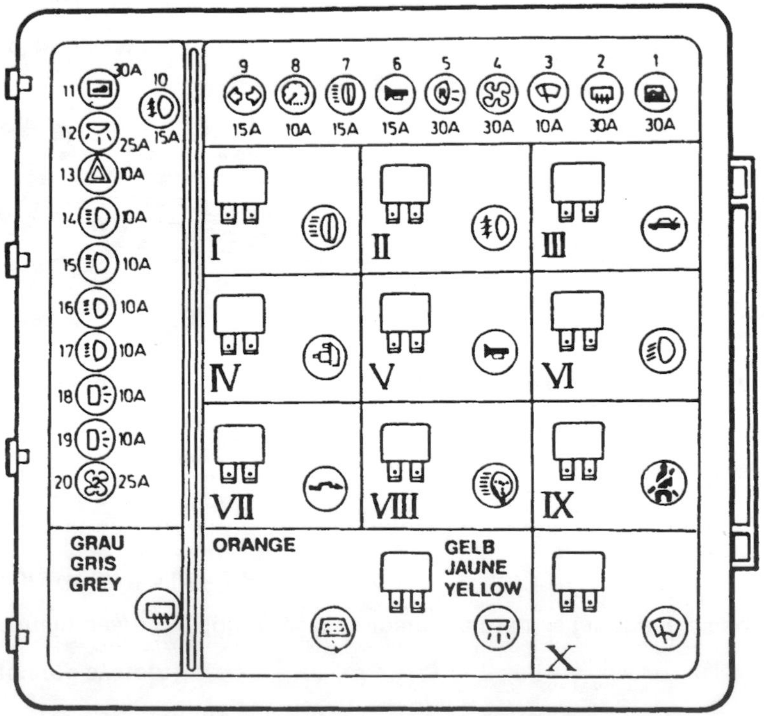

CIRCUITS OPERATED BY RELAYS	
i	viii
ii	ix
iii	x
iv	Grey
v	Yellow
vi	Orange
vii	

MAIN CIRCUITS PROTECTED BY FUSE	
1	11
2	12
3	13
4	14
5	15
6	16
7	17
8	18
9	19
10	20

State what the following symbols represent:

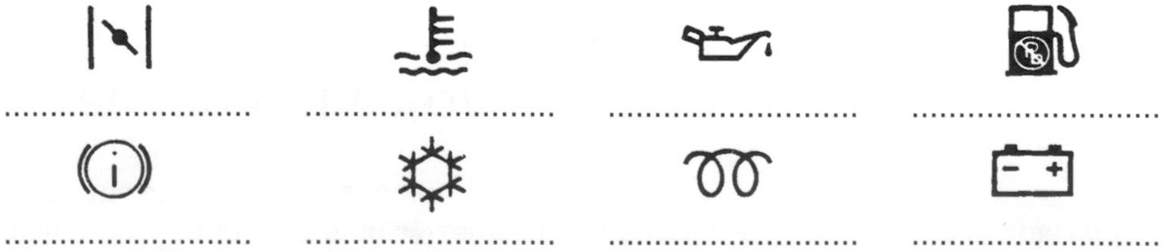

The diagram illustrates the instrument panel of a Ford Granada 2.0i Ghia.

Enlarged are shown 40 standard symbols which are positioned on the vehicle instruments and control knobs.

Examine the position of these items and identify their function using the similar drawings on the next page.

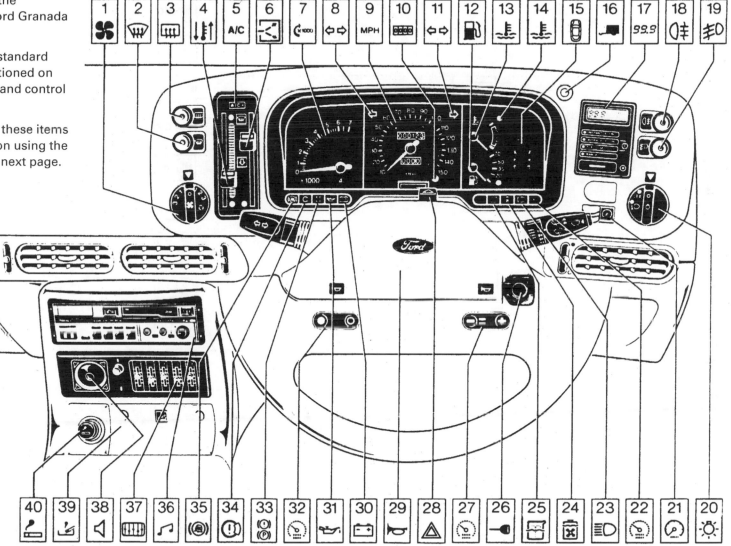

Name the instrument/control knob symbols shown below:

1 �含 ....

2 ▨ ....

3 ▨ ....

4 ↓▮↑ ....

5 A/C ....

6 ◁ ....

7 ⊂⋯ ....

8 ⇦⇨ ....

9 MPH ....

10 ▦ ....

11 ⇦⇨ ....

12 ▐ ....

13 ▟ ....

14 ▟ ....

15 ▯ ....

16 ◢ ....

17 99.9 ....

18 ◖‡ ....

19 ⹋◗ ....

20 ☀ ....

21 ◔ ....

22 ◕ ....

23 ▤◗ ....

24 ▨ ....

25 ▢ ....

26 ◀ ....

27 ◔ ....

28 △ ....

29 ▷ ....

30 ▭ ....

31 ▱ ....

32 ◕ ....

33 ▨ ....

34 ⊕ ....

35 ◉ ....

36 ♪ ....

37 ▦ ....

38 ◁ ....

39 ◣ ....

40 ◢ ....

CIRCUIT COMPONENTS

The components shown on this page are used in various motor-vehicle electrical circuits.
Identify the components and state their basic function

Sketch the components' wiring symbols.

Electrical Components

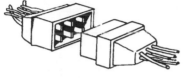 

Electronic Components

Sketch the components' wiring symbols.

 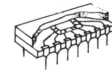

ELECTRONIC SYSTEMS ON VEHICLES

Identify the electrical/electronic components shown on this page and state what they control.

Ignition

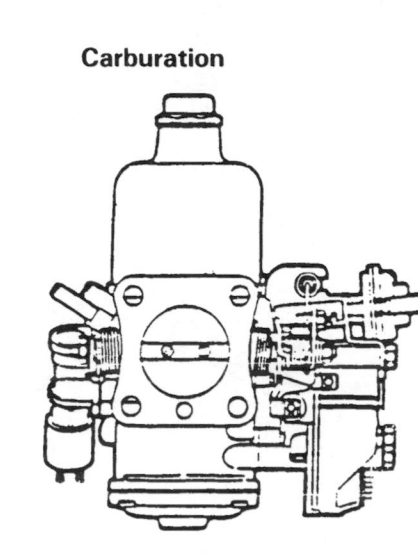

....

....

Carburation

....

....

Alternator

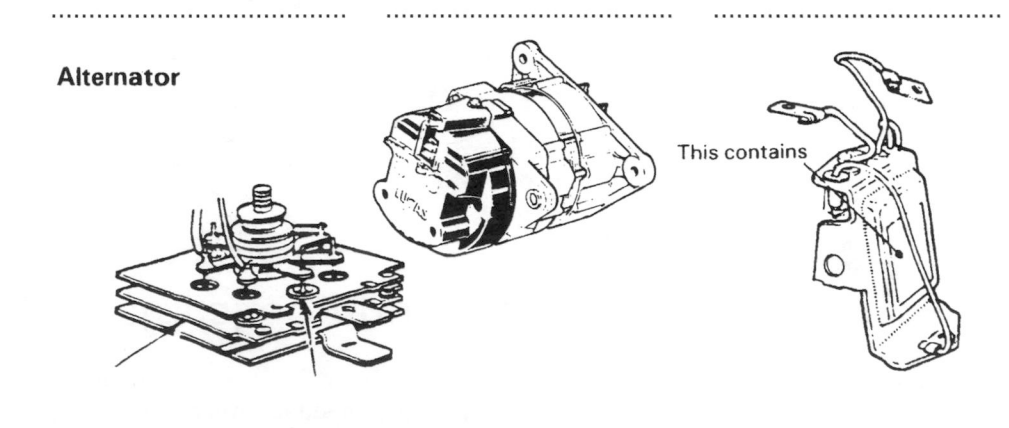

This contains

....

....

Instrumentation

Name the items on the drawing that are controlled electronically.

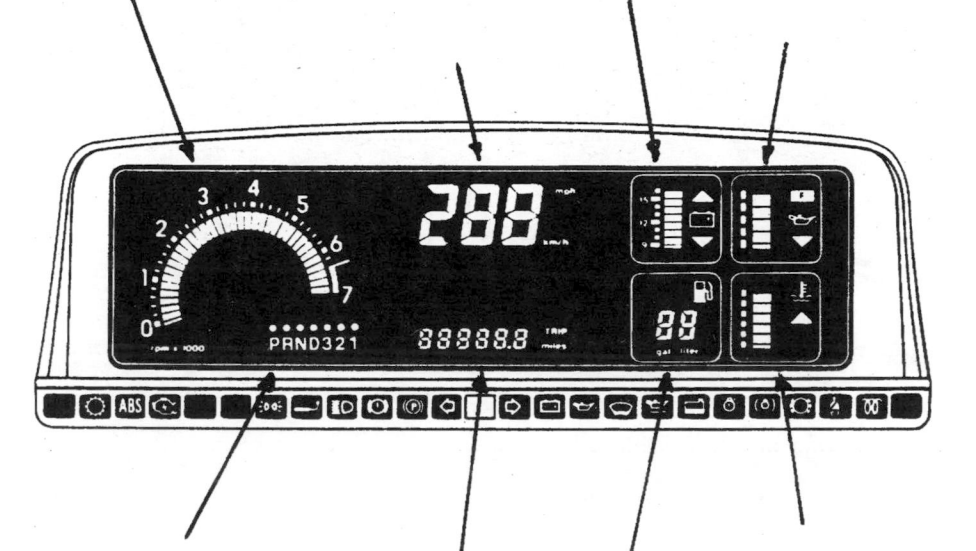

Fault finding is also carried out using electronic diagnostic equipment. One example of this is the use of an oscilloscope for checking the ignition system. Below are three different traces obtained by an oscilloscope.
State what each trace indicates.

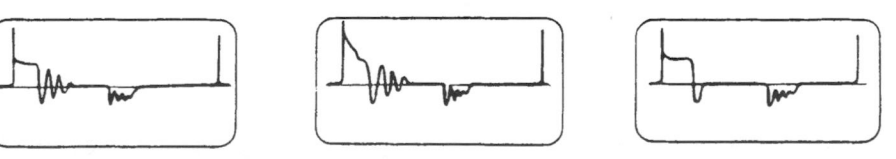

....

....

ELECTRONIC SYSTEMS ON VEHICLES

All electronic systems use different forms of sensors to supply an electronic control unit (ECU) with information. The ECU will first read the information, then compare it with programmed information in its memory; it will then send a signal to an actuator to alter the control situation if required. Depending on type, the system may control the engine fuel system, ignition system, automatic transmission system, ABS, etc.

The basic design for all electronic systems is called an 'open loop' system.

Name the items represented by the open loop shown and state their function.

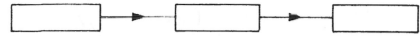

A practical example would be the electronic ignition system.

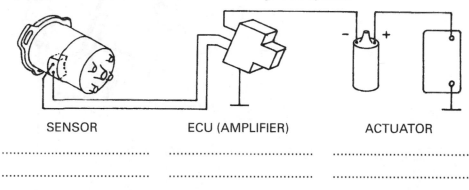

SENSOR ECU (AMPLIFIER) ACTUATOR

Complete the open loop control diagram below to show how fuel supply in an electronic fuel injection system is cut off when the engine speed is over 2000 rev/min and the throttle is closed. That is, the vehicle is slowing down from a high speed, engine on overrun.

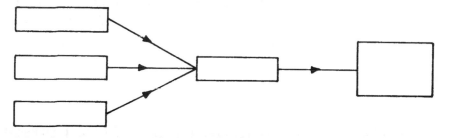

Engine Electronic Control Unit

Electronic circuits operate on very low voltages, and to ensure long-term high-quality terminal connections, special types of contracts and plugs have been introduced. How should the multi-pin plug be handled?

What precautions should be taken before disconnecting the plug from an ECU?

Describe how the plug should be removed and replaced.

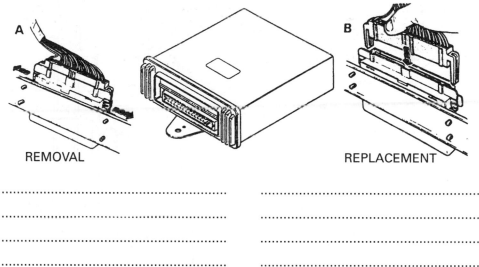

A

REMOVAL

B

REPLACEMENT

123

Bosch LE Jetronic Petrol Injection System

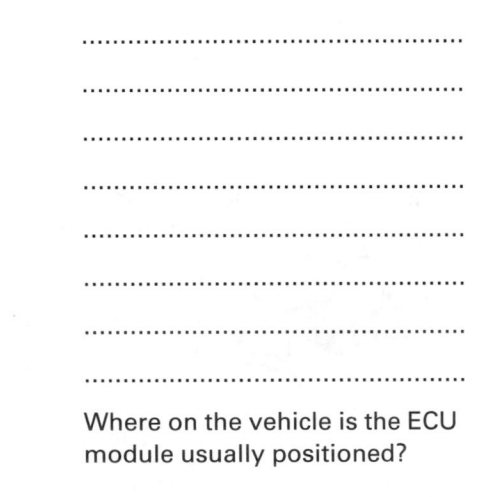

The diagram shows an electronic control unit (ECU) and the sensors and actuators to which it is connected.

Describe the ECU's function.

....
....
....
....
....
....
....
....
....

Where on the vehicle is the ECU module usually positioned?

....
....
....
....
....

Name below the parts numbered on the drawing of the wiring circuit and indicate the direction of air and fuel flow.

FUEL SYSTEM

1. ....

2. ....

3. ....

4. ....

5. ....

6. ....

COLD START SYSTEM

7. ....

8. ....

SENSOR SYSTEM

9. ....

10. ....

11. ....

12. ....

13. ....

14. ....

CONTROL SYSTEM

15. ....

16. ....

Electrical Test Equipment

Ammeter

The ammeter measures current flow in amperes. How is an ammeter connected into the circuit to measure current flow?

....

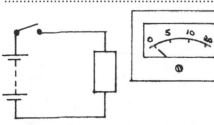

Draw an ammeter in position on the circuit shown above.

Voltmeter

The voltmeter measures electrical pressure in volts. How is a voltmeter connected into the circuit?

....

Draw a voltmeter in position on the circuit shown above.

What types of ammeter and voltmeter are shown above?

....

When testing electronic circuits meters having a very high input impedance (resistance) are preferred. These are digital meters.

Ohm-meter

The ohm-meter looks similar to the two other meters shown above; it measures electrical resistance of circuits or components in circuits. What is a feature of the ohm-meter which does not occur with the other types of meter?

....
....
....
....

What precautions must be taken before checking a component with an ammeter?

....
....

Multimeters

All multimeters will measure voltage (both a.c. and d.c.), current and resistance.

Examine the type of meter used in the workshop, and sketch its face in the block opposite, naming the important features.

Meter   Name/number

....
....

Below is shown a multimeter connected to electronic equipment. State to which meter the settings will be connected and the type of test being carried out.

....
....

....
....

ELECTRICAL TESTING

Describe how the resistance of a component should be tested using an ohm-meter or multimeter.

1. ....

2. ....
....

3. ....
....

Select FIVE electrical/electronic components and measure their resistance.

COMPONENTS					
RESISTANCE					

How can current flow through a circuit be tested?

....
....

Describe how a voltmeter should be used to check the continuity of a circuit for high resistance.

....
....
....
....

Which meter may be used to check the continuity of a non-live circuit? ....

PRECAUTIONS WHEN TESTING OR REPLACING COMPONENTS

Describe any special precautions to be observed when:

1. Disconnecting electrical components ....
....
....
....
....
....

2. Replacing components/units ....
....
....
....
....
....
....

3. Testing components ....
....
....
....
....

What faults are especially dangerous to electronic components?

1. ....

2. ....
....

BUILDING VARIOUS ELECTRICAL CIRCUITS

Equipment:

various types of resistances; switches; cables; ammeter; voltmeter; battery

Assemble the circuits described below; include in each circuit a switch and battery. In at least TWO circuits show the correct connection of an ammeter.

You may be provided with *either*

(a) motor-vehicle components   *or*   (b) a peg board   *or*   (c) a construction kit.

Sketch the circuit diagrams as required and state the total current flow in each case.

In a series circuit, the bulbs or resistances are connected:

....

....

....

In a parallel circuit, the bulbs or resistances are connected:

....

....

....

---

1. Two resistances in parallel

Current flow ....

2. Four resistances connected in series

Current flow ....

3. Six resistances connected in parallel

Current flow ....

---

4. Two resistances in series connected with one resistance in parallel

Current flow ....

5. Two sets of two resistances in parallel connected to one resistance in series. Both sets then being connected in parallel with one another.

Current flow ....

127

SERIES AND PARALLEL CIRCUITS

Build two electrical circuits as shown.

Series circuit

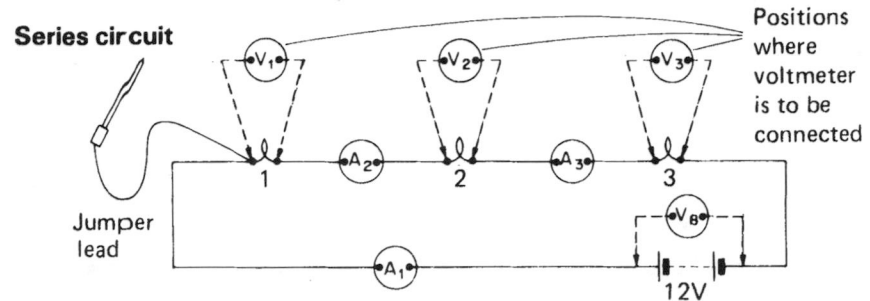

Parallel circuit

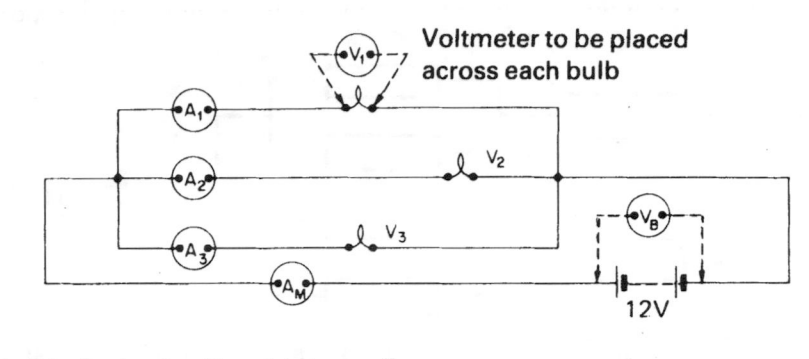

Connect all bulbs in circuit and note readings:

Voltage

V_1	V_2	V_3	Total	V_B

Current flow

A_1	A_2	A_3

Why are the bulbs dim? ....

Remove bulb 1, what happens? ....

Why does this occur? ....
Place prod of jumper lead to bulb 2, take readings:

Voltage

V_2	V_3	Total	V_B

Current flow

A_1	A_3

What differences have now occurred and why?

....

Remove bulb 2, connect prod of jumper lead to bulb 3, take readings:

Voltage

V_3	V_B

Current flow

A_1

Connect all bulbs in circuit and note readings:

Voltage

V_1	V_2	V_3	V_B

Current flow

A_1	A_2	A_3	Total	A_M

When three bulbs are connected, compared with the series circuit what differences occur with regard to the following? Give reasons.

Voltage ....

....

Current flow ....

....

Light intensity ....

....

Note current flow when bulbs are removed:

	A_1	A_2	A_3	Total	A_M
One bulb removed					
Two bulbs removed					

Why did the same thing not occur as in the series circuit?

....

128

CURRENT, VOLTAGE, RESISTANCE – OHM'S LAW

Ohm's Law is the expression that relates voltage, current and resistance to each other.

State what these electrical terms mean, and state units in which they are measured.

Voltage ....

....

Current ....

....

Resistance ....

....

One of the relationships defining Ohm's Law states that, provided the resistance is kept constant, the current will double if the voltage is doubled. Expressed more mathematically this could be stated as:

....

....

Expressed as a formula using electric symbols

$$I = \frac{V}{R}$$

where I = ...

R = ...

V = ...

The formula may be transposed to state

$V =$ .... and $R =$ ....

Example

Calculate the current flowing in a circuit when a pressure of 12 V is applied across a 3-ohm resistance

Simple Ohm's Law Problems

1. Calculate the current flowing in a coil of resistance 4 ohms when the electrical pressure is 12 V.

   ....

   ....

   ....

   ....

   ....

2. Calculate the voltage required to force a current of 2.5 amps through a resistor of 5 ohms.

   ....

   ....

   ....

   ....

3. An alternator produces a current of 35 A when the voltage is 14 V. What is the resistance of the alternator?

   ....

   ....

   ....

   ....

4. What voltage will be required to cause a current flow of 3 A through a bulb having a filament resistance of 4.2 ohms?

   ....

   ....

   ....

   ....

5. What will be the total resistance offered by a lighting circuit if a current of 11 A flows under a pressure of 13 V?

   ....

   ....

   ....

   ....

6. Two 12 V headlamp bulbs each have a resistance of 2.4 ohms. Calculate the current flowing in each bulb and the total current flowing in the circuit.

   ....

   ....

   ....

   ....

SERIES CIRCUITS

The basic laws of series circuits are stated below. Show calculations to obtain values when the battery voltage and resistances are as shown.

The ammeters (diagram 2) and voltmeters (diagram 3) are shown in their relative testing positions.

1. RESISTANCE   The resistance of a series circuit is the sum of each separate resistance in the circuit.

TOTAL R = ...

....

2. CURRENT   The current in a series circuit is the same in all parts of the circuit. To find total current use Ohm's Law.

TOTAL
CURRENT = ...

....

3. VOLTAGE   The voltage of a series circuit is the sum of the voltages across each separate resistor.

Using Ohm's Law $V = I \times R$

....

....

....

....

1. Three resistors of 15, 30 and 12 Ω are connected in series. What is the total resistance in the circuit?

....

....

....

2. Five resistors of 100 Ω, 500 Ω, 1 kΩ, 1.5 kΩ and 1 MΩ are connected in series. What is the total resistance of the circuit?

....

....

....

3. Four resistors of 6, 10, 14 and 18 Ω are connected in series to a 12 V circuit. Calculate the total resistance of the circuit and the current flowing in each resistor.

....

....

....

....

....

....

4. Two resistors of 1.25 Ω and 3.75 Ω are connected in series. What voltage would be required to cause a current of 2.5 A to flow in the circuit?

....

....

....

....

....

5. Three resistors of 2, 4 and 6 kΩ are connected in series to a 12 V battery. Calculate the current flowing in the circuit.

....

....

....

....

....

6. Two resistors each of 2 kΩ are connected in series to a 9 V supply. Calculate the current flowing and the voltage between the two resistors.

....

....

....

....

PARALLEL CIRCUITS

The basic laws of parallel circuits are stated below. Show calculations to obtain values when the battery voltage and resistances are as shown.

The ammeters (diagram 2) and voltmeters (diagram 3) are shown in their relative testing positions.

1. RESISTANCE

Total or equivalent resistance R is obtained from the formula:

$$\frac{1}{R} = \frac{1}{R_1} + \frac{1}{R_2} + \frac{1}{R_3}$$

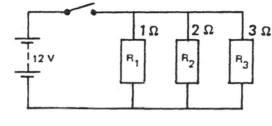

....

....

....

....

The total resistance of a parallel circuit is always less than the smallest resistance in the circuit. The more branches in the circuit the easier it is for the current to return to the battery.

2. CURRENT

Current through a parallel circuit is the sum of the current through each separate branch of the circuit.

TOTAL I = ...

3. VOLTAGE

Voltage of a parallel circuit is the same across each separate resistor.

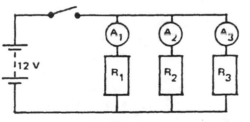

VOLTAGE = ...

Problems

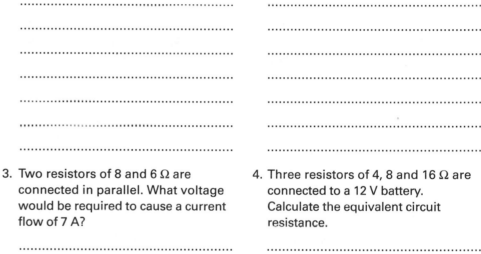

1. Two resistors of 3 and 4 Ω are connected in a 12 V parallel circuit. Calculate the total current flow.

....

....

....

....

....

....

....

2. Three conductors are placed in a parallel circuit, their resistances being 4, 8 and 12 Ω. Calculate the current flow in each resistor and the total current flow when connected to a 12 V system.

....

....

....

....

....

....

....

3. Two resistors of 8 and 6 Ω are connected in parallel. What voltage would be required to cause a current flow of 7 A?

....

....

....

....

....

....

....

....

....

4. Three resistors of 4, 8 and 16 Ω are connected to a 12 V battery. Calculate the equivalent circuit resistance.

....

....

....

....

....

TOTAL I = ...

....

....

....

....

Which circuit – series or parallel – divides its total:

current? .... resistance? .... voltage? ....

The division of voltage is most important, particularly in electronics, since a varying voltage allows the switching on and off of components requiring control.

Potential Divider

This divides potential (voltage) in a series circuit. If a variable resistor is used, the output voltage can be varied and provide control.

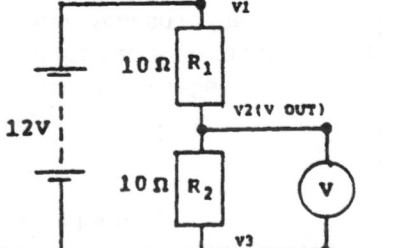

From previous work, if the resistances are equal then voltage at V = half the total voltage; this is known as the output voltage or voltage out. By altering the values of the resistors, this voltage can be varied to give very high or very low values.

From diagram, Voltage out = ...

VOLTAGE OUT can be calculated using the formula

$$\text{Voltage out} = \frac{R_2}{R_1 + R_2} \times \text{supply voltage}$$

Calculate the voltage out values when:

1. $R_1 = 1\,\Omega$ and $R_2 = 59\,\Omega$

2. $R_1 = 100\,\Omega$ and $R_2 = 5\,\Omega$

ELECTRICAL POWER

Power is the rate of doing work and is measured in ....

The definition of electrical power is: ....

The power consumption in electronic circuits is very small, milliwatts (mW) or microwatts (µW), whereas the power consumed by lamps may be measured in watts (W), and by starter motors in kilowatts (kW).

1. If an alternator is generating a current of 60 A with an output voltage of 14 V, what power is required to turn the alternator?

2. A starter motor when cranking an engine consumes a current of 135 A. The terminal voltage is 11.5 V. Calculate the starter motor's power output.

3. Calculate the amount of current used by a 12 V, 48 W headlamp bulb.

4. Calculate the amount of current used by two 12 V, 54 W headlamp bulbs.

Voltage Drop

In electrical systems the voltage (or pressure) across the component being operated – say the starter motor – should ideally be the same as the battery voltage. However, in practice it is usually less. This is because the copper battery cable has a small amount of resistance. Therefore it will have a small voltage drop across it when current flows to the starter motor. If the battery leads have too high a resistance then the voltage drop will be too high for the starter motor to work effectively.

Define the following terms:

e.m.f. ....

....

p.d. ....

....

v.d. ....

....

The voltage drop in a circuit   v.d. =

EXAMPLE

During starter operation the battery voltage is measured to be 10.5 V. The voltage measured at the battery terminal of the solenoid is 9 V; determine the v.d.

v.d. = ...

Alternatively if the resistance and the current flow of a component are known, the v.d. may be found by applying Ohm's Law.

$$V = I \times R$$

EXAMPLE

Calculate the v.d. in a starter cable when the internal resistance of the cable is 0.0012 Ω when carrying a current of 220 A.

$V =$ ....

Problems

1. A battery has a p.d. of 9.75 V when operating a starter motor. The open circuit battery e.m.f. is 12.5 V. Determine the v.d. when the starter is in operation.

....

....

....

....

2. If the volt drop across a starter cable is 2.75 V and the e.m.f. is 12.85 V, determine the terminal voltage at the starter motor when it is in operation.

....

....

....

....

....

3. Calculate the v.d. across a starter cable if the resistance of the cable is 0.0016 Ω when carrying a current of 284 A.

....

....

....

....

4. A 12 V battery gave an open-circuit voltage of 12.77 V. When supplying a current of 18 A the terminal voltage drops to 12.05 V. Calculate the internal resistance of the battery.

....

....

....

....

....

....

5. The leads from a 12 V battery to a starter motor have a total resistance of 0.005 Ω. What p.d. will be required to send a current of 200 A through those leads? What would be the voltage at the starter motor terminals in this case?

....

....

....

....

....

....

RESISTANCE OF ELECTRICAL CABLES

If the length of cable supplying current to a component is doubled, the internal resistance of the cable

will ....

that is, length is directly ....

.... to the resistance.

Problems

1. If the resistance of a wire is 0.009 Ω when it is 6 m long, what will be its resistance when it is 72 m long?

   ....

   ....

   ....

2. If 5 m of starter cable has a resistance of 0.025 Ω, calculate the resistance of 30 m of this cable.

   ....

   ....

   ....

3. Calculate the resistance of 1.5 m of cable if 60 m has a resistance of 0.05 Ω.

   ....

   ....

   ....

   ....

If the cable's cross-sectional area is doubled, the internal resistance will

....

that is, cross-sectional area is ....

.... to the resistance.

Problems

1. A cable has a resistance of 0.04 Ω and an area of 15 mm². If the area is increased to 90 mm², what would be the resistance?

   ....

   ....

   ....

2. A cable has an area of 2 mm² and a resistance of 0.03 Ω. If the area is increased to 72 mm², what would be the resistance?

   ....

   ....

   ....

3. A cable has a cross-sectional area of 105 mm² and a resistance of 0.005 Ω. What would be the resistance if the area was only 7 mm²?

   ....

   ....

   ....

SHORT AND OPEN CIRCUITS

Two simple light circuits are shown, one is an open circuit, the other has a short circuit.
Identify, giving reasons for choice:

The circuit ....

....

....

The circuit ....

....

....

What could be a probable cause and effect of the short circuit?

....

....

....

....

A common fault of electrical circuits is unwanted series resistance. What is meant by this term and how does the resistance occur?

....

....

....

....

....

....

....

....

Chapter 5

Vehicle Valeting
(Level 1)

BASIC MATERIALS AND TERMINOLOGY

Professional valeting today is a far cry from the simple bucket and sponge, although they still have their uses. There is now a vast range of specialist preparations, each designed to do a particular job in the most efficient possible manner. List below TEN distinct types of valeting preparations:

1. *New car dewaxent* ....
2. ....
3. ....
4. ....
5. ....
6. ....
7. ....
8. ....
9. ....
10. ....

Inevitably these preparations need to be used with care, especially as they could well contain a 'cocktail' of possibly hazardous chemicals.

From where can basic information regarding their proper use and appropriate safety precautions be easily obtained?

....

TYPICAL HAZARDOUS PRODUCT LABEL

(According to the Classification, Packaging and Labelling of Dangerous Substances Regulations 1984)

Study this label and state the significance of the indicated sections:

NIELSEN DISCO WHEEL CLEAN

Alloy Wheel Cleaner

Nielsens Disco Wheel Cleaner is a complex blend of acids and surfactants which is highly effective in removing black carbon dust and dirt from alloy and painted wheels.

DIRECTIONS FOR USE: Spray Disco Wheel Clean onto wheel using the Nielsens Wheel Cleaning Sprayer. Leave for 3 minutes and then pressure wash off using hot or cold water. Does not affect rubber.

HEALTH & SAFETY AT WORK ACT 1974
NIELSEN CLASSIFICATION D
DISCO WHEEL CLEAN: Contains 12% Phosphoric Acid. Substance Identification No. 1805.

HANDLING PRECAUTIONS: Irritating to eyes and skin. Avoid contact with eyes and skin.

FIRST AID: Eyes: Rinse immediately with plenty of water and seek medical advice. Skin: Wash off with soap/water. Ingestion: Rinse out mouth frequently with water and seek immediate medical advice.

STORAGE: Store in containers provided. Temperature not to exceed 40°C.

SPILLAGE: Rinse affected area thoroughly with plenty of water.

FIRE: Non-flammable.

IRRITANT

Nielsen Chemicals Limited,
Stanhope Road, Swadlincote, Derbyshire, DE11 9BE.
Telephone: (0283) 221044. Fax: (0283) 225731.

Look at the labels on a number of containers. Several terms frequently appear. Some of the most common are mentioned below. State briefly what they mean.

SOLVENT AND SOLUTE

....

....

....

SOLUBLE AND INSOLUBLE

....

....

....

What is the most common solvent using in valeting?

....

Name another common solvent, but one made from hydrocarbons (petroleum distillates):

....

Solvents must be appropriate to the substances with which they are to be mixed. For example, to clean brushes used for painting tyres, it would be foolish to use water – it would not act as a solvent for the petroleum-based paint. (White spirit is the probable solvent.)

There is a very simple 'general rule' regarding solvents, which is:

....

Valeting preparations may be acids or alkalis. Both of these can be dangerous. Strong concentrations can cause:

On metals and paint ....

On skin or eye contact ....

What hazard do all petroleum-based products have in common?

....

DETERGENTS, SYNDETS AND SURFACTANTS

A DETERGENT is ....

What are the TWO most common groups of detergent?

....

Ordinary soaps are made from organic (natural) materials such as animal fats and vegetable oils. These are know as ESTERS and are BIODEGRADABLE. This means:

....

What may be one practical problem using soaps with hard water?

....

SYNDETS are soapless SYNthetic DETergents and are produced by the petroleum industry.

One prime advantage of syndets is ....

....

In the past, syndets were not biodegradable and caused masses of foam on rivers. However, the latest improved formulations are biodegradable.

SURFACTANTS: Both soaps and syndets are SURFACTANTS. It means that they clean the surface. The name is made up from:

....

Great care must be taken in the selection of surfactants for any particular job. For example, one made specially for plastic, if used on glass could:

....

Detergents are normally combined with a DISPERSANT. Their combined action is to clean the surface and hold the dirt in very finely divided particles that can be easily rinsed off in water.

SAFETY PRECAUTIONS

It is essential to choose the correct products to carry out valeting operations quickly and effectively, but also SAFELY. The regulations, Care of Substances Hazardous to Health (COSHH), require manufacturers of these products to make available – on request – more detailed information than that shown on the product container.

PRODUCT SAFETY DATA SHEET

An example of such a data sheet is shown opposite. Employers must obtain this information and make its location and availability known to employees.

What, briefly, are the obligations of an employee?

....

....

....

Study the example of a product Safety Data Sheet (opposite) and answer the questions below.

Purpose of product ....

How should it be applied? ....

....

Name one of the hazardous ingredients ....

Name one material that reacts with it ....

What would be the resulting danger of the reaction? ....

....

How should such a fire be treated?

....

How should small spillages be dealt with?

....

Product Safety Data Sheet

Product: HEAVY DUTY FABRIC CLEANER   Product Code: 214

SECTION 1:   PRODUCT DESCRIPTION & PHYSICO CHEMICAL DATA

APPLICATION: Detergent Concentrate for use through hot water soil extraction carpet/fabric seat cleaning machines.

PHYSICAL FORM: Water thin clear blue liquid, citrus odour.

CHEMICAL COMPOSITION: Sequestering agents, surface active agents, sodium hydroxide, 1-Methoxpropan-2-ol, hydrotope, optical brightener, Perfume oils, dye, preservatives, water.

HAZARDOUS INGREDIENTS:

Material	Nature of Hazard
Sodium Hydroxide	Causes damage to eyes, skin and clothes.
1-Methoxypropan-2-ol	Maximum exposure limit set (M.E.L.) (see below). Can be absorbed through skin.
M.E.L. = 1-Methoxypropan-2-ol =	8 hr TWA: 100 ppm 10 min TWA: not set

SECTION 2:   FIRE/EXPLOSION/REACTIVITY DATA

Product is non-flammable. However, irritating fumes may be given off in the event of a fire. Treat fires with dry chemical, foam or waterspray.

Reactivity: Alkaline products react with light metals (such as aluminium, tin or zinc), with the evolution of hydrogen gas which is highly flammable.

SECTION 3:   HEALTH HAZARD INFORMATION

Personal protection/precautions in use. Use only according to instructions. Always wear impervious gloves and eye protection when handling this chemical. Do not mix with other chemicals. Always maintain a good standard of occupational hygiene when handling chemicals.

Effects of over exposure

Skin contact: Degreases skin, may lead to redness and cracking. Irritation will be felt on sensitive skin. Prolonged contact causes damage.

First Aid: Remove contaminated clothing, wash affected area with soap and water.

Eye contact: Severe pain, redness, watering. May cause permanent damage.

First Aid: Wash chemical out of eyes immediately. Use plenty of clean water for at least 15 minutes. Seek immediate medical attention.

Ingestion: Sore mouth and throat, abdominal pain, vomiting.

First Aid: Rinse mouth out with water then drink milk or water and seek immediate medical attention.

Inhalation: Mild irritation.

First Aid: Move casualty to fresh air, seek medical attention if necessary.

SECTION 4:   STORAGE

Store away from children. Always ensure cap is tightened after use. Keep away from food stuffs.

MAX/MIN TEMPERATURES: 0°C–35°C

SECTION 5:   SPILLAGE/LEAK PROCEDURE/WASTE DISPOSAL

Small spillages can be rinsed away to drain. Large spillages should be contained and advice sought from Local Water Company.

Disposal: According to local regulations.

VALETING INSTRUCTIONS

As at any other time when you are in contact with a customer, it is important that you project a favourable image of your firm or organisation. One way of doing this is to listen carefully to what the customer tells you and accurately record his/her instructions and requirements.

Preferably this should be done on a printed form which is completed in the customer's presence. It should be signed by the customer, who should then be given a copy.

A typical layout of a Valeting Instruction Sheet is shown opposite. This sort of form, properly completed, may be used for evidence in your portfolio for N/SVQ purposes.

Why indicate marks or damage prior to valeting? ....

....

Although not specifically mentioned on the form, the valeter must be watchful for various other faults that should be brought to the attention of the customer. Give FOUR examples:

1. ....

2. ....

3. ....

4. ....

Why is the customer also asked to sign the form on collection?

....

....

What should be done regarding personal items, loose tools, etc. that have been left in the car?

1. ....

2. ....

VALETING INSTRUCTION SHEET

Customer Name:

Address: Phone: Business
                                                                                                         Home

Vehicle Details: Vehicle
Make   Model   Reg. No.:
Type of Valeting Operation Required:

PDI   Full Valet   Exterior Only   Interior Only   Engine Bay

Special Instructions ....

....

On completion, Protective Covers to be left: ON   OFF

Record of marks/damage prior to valet service

Exterior:

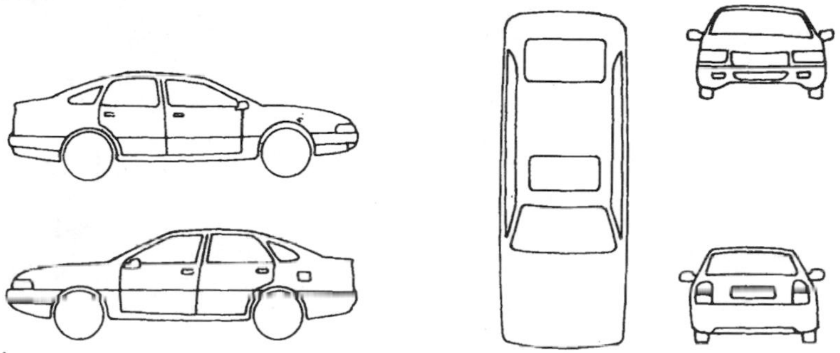

Interior:

Time vehicle required: Customer signature to confirm instructions

                                                    ....

On completion, the customer should sign that the above instructions have been satisfactorily completed.

Customer signature .... Date ....

Valet service carried out by ....

139

PERSONAL PROTECTIVE EQUIPMENT (PPE)

To protect the operator during valeting work, especially on the exterior, appropriate protective clothing and equipment must be used – plus observing all necessary safe practices (e.g. Local by-laws and Environmental Protection Act).

The provision and use of PPE is required by the European Committee for Standardisation (often referred to by the initials CEN). There are two main DIRECTIVES. They concern:

1. ....   2. ....

The symbol shown appears on many items of PPE. What is it called and what is its purpose?

CE

....
....
....
....

Do employees have to purchase their own PPE?

....
....

List FOUR forms of PPE that may be required by anyone continuously engaged in cleaning vehicle exteriors:

1. ....
2. ....
3. ....
4. ....

These and other safety requirements are embodied in the United Kingdom, Health and Safety at Work Act, of which the COSHH regulations, previously mentioned, are a part.

Typical Personal Protective Equipment (Exterior Cleaning)

Name the protective equipment items represented below by their symbols, and alongside list some of their main features.

....
....
....

....
....
....

....
....
....

....
....
....

WASTE DISPOSAL

It is most important that the disposal of waste, inevitable with the valeting process, is dealt with safely, in accordance with statutory and organisational requirements.

The majority of such waste is 'dirty' water from the various washing procedures. List THREE main contaminants:

1. ....

2. ....

3. ....

The drawing below shows a typical garage drain system that might be used with a valeting operation. On it, indicate the water path and label the main parts.

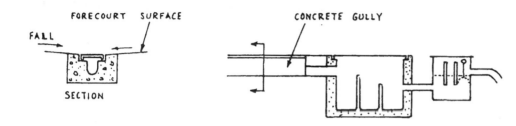

Why is such a system necessary?

....

....

....

....

....

What precautions are necessary with used aerosols?

....

....

VALETING OPERATIONS

The extent of any valeting operations can vary considerably. List FOUR factors that are likely to affect them:

1. ....

2. ....

3. ....

4. ....

General valeting operations can be divided into THREE areas:

1. ....

2. ....

3. ....

It is usual in a professional valeting operation that the three operations would be carried out by different staff. Why is this?

1. ....

2. ....

3. ....

Protective Covers

Where on a vehicle might protective covers be used during valeting?

1. ....

2. ....

3. ....

4. ....

5. ....

NEW CAR DEWAX

To safeguard them while in transit from factory to customer, many new cars are given a protective coating of wax or lacquer – usually referred to as HARD or SOFT wax. This must be removed prior to delivery. Suggest a typical procedure.

First, determine, the TYPE of wax used, so as to select the correct solvent.

SOFT WAX ....

....

HARD WAX ....

....

State ONE appropriate solvent for EACH type of wax:

SOFT WAX ....

HARD WAX ....

Typical Procedure – Soft Wax (using pressure washer)

1. *Observe all necessary safety precautions*

   ...... *and product instructions* ....

2. ....

   ....

3. ....

   ....

4. ....

   ....

5. ....

   ....

6. ....

   ....

7. ....

   ....

8. ....

   ....

9. ....

   ....

What are THREE main differences when dealing with HARD WAX?

1. ....

   ....

2. ....

   ....

3. ....

   ....

Inspect several new cars 'in wax' and determine with what type of wax they are covered.

MAKE	MODEL	WAX TYPE
....	....	....
....	....	....
....	....	....
....	....	....

EXTERIOR VALETING (late used car)   Complete this brief guide

WHERE ON VEHICLE	Vehicle exterior	Alloy wheels	Body work	Glass & mirrors	Plastic trim	Tyres
PRODUCT TYPE	Traffic Film Remover (TFR)					Rubber Dressing
COMMERCIAL PRODUCT	Nielsen Transclean 2000					Nielsen Rapport
HOW TO USE IT	Via Pressure washer. 100:1 Hot 10:1 Cold					Apply via hand spray
SAFETY	Irritant		Flammable	Flammable		Not COSHH Classified

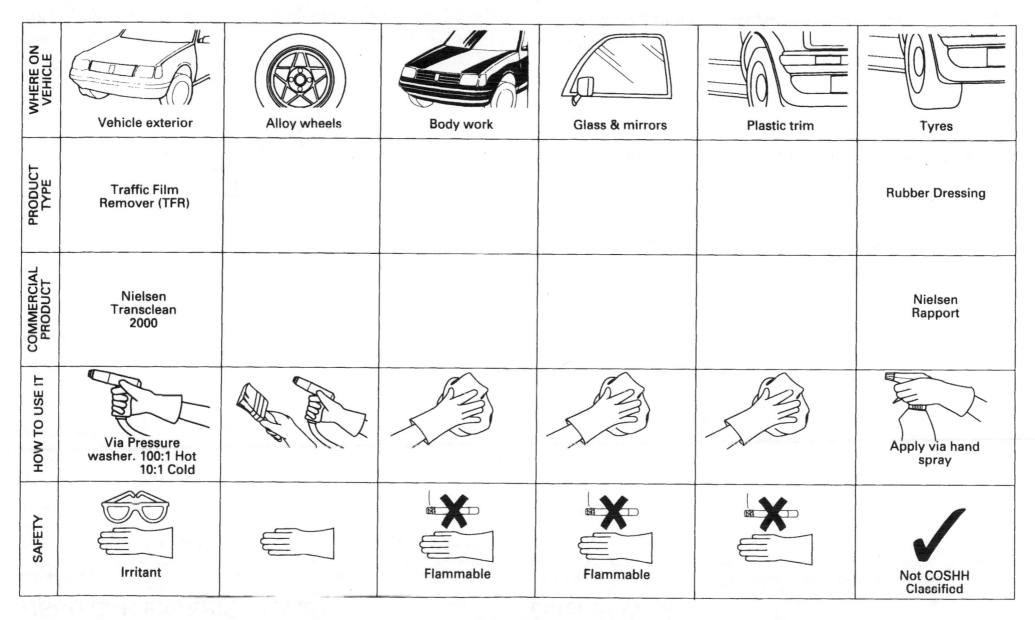

USED CAR EXTERIOR VALET

Reinstating the cleanliness of a relatively new (late used) vehicle in good order is generally much less difficult than with an older used vehicle. However, it is with the older and maybe more badly used vehicles that the valeter can show the best results. Describe some of the major points listed below.

SAFETY ....

....

WHEELS ....

WHEEL ARCHES ....

....

ENGINE ....

....

DOOR SHUTS, ENGINE, BOOT SURROUNDS etc. ....

....

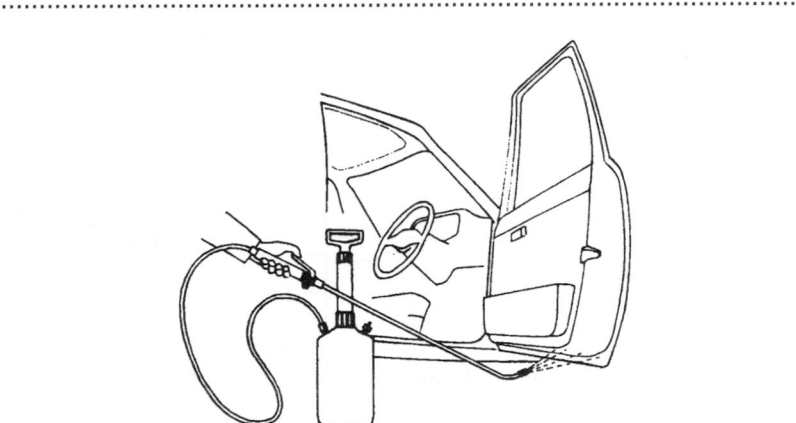

HEAVILY SOILED AREAS, e.g. bonnet underside, door hinges ....

....

TFR-REMOVAL ....

....

Exterior Wash

After TFR-removal, the car must be washed. Many alternative systems and preferences exist. Mention one that you consider gives good results.

SHAMPOO ....

....

RINSE ....

....

Tar Deposits on Paintwork

Name ONE proprietary brand and ONE common alternative:

....

....

Restoring Aged or Faded Paintwork

Polishes used for this purpose have a mildly abrasive or cutting action. Name TWO widely used such products:

....

....

Such polishes cut away the chalky type deposits and road deposits on the dead surface of the old paint, so as to restore its colour and sheen. The abrasive qualities of different polishes vary. It is important to realise that some paints are harder (more abrasive-resistant) than others. In general, how may they be distinguished?

'SOFT' PAINTS ....

'HARD' PAINTS ....

TWO-PACK METALLIC PAINT – ensure that polish is compatible.

Typical two-pack polish ....

What are TWO other uses of abrasive type polishes?

....

Opinions vary as to how polish should be applied and then buffed off. It is though *generally* agreed that:

When applying polish, pressure should be ....

When buffing, pressure should be ....

For speedy application of polish and buffing, it is common to use air or electrically powered mops.

Two types of mop heads are:

Polyurethane foam ....

Lambswool ....

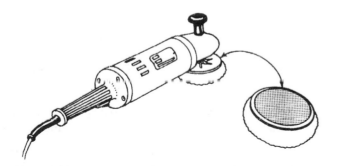

How is the mop attached to head?

....

Certain areas should not be treated by machine for fear of burning through paint surface. For example:

....

Polishing and Buffing

Both during polish application and buffing, the direction of hand movement is important. See the sketches below and label them to indicate what shows polish application and what shows buffing. Sketch the direction of application on other panels.

COMMONLY USED METHOD

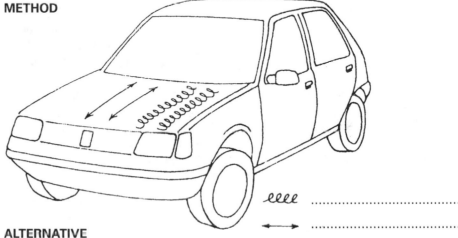

eeee ....

↔ ....

ALTERNATIVE METHOD

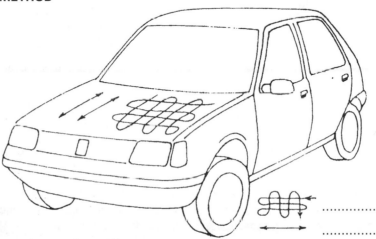

....

↔ ....

WAX POLISHES

The brightness of paintwork newly cleaned by use of a 'cutting' process can quite quickly fade. How should it be preserved?

....

....

The application of wax polish tends to fill in the tiny scratches of the newly cleaned paintwork and give a smooth, weather-resistant surface which reflects the light and enhances its appearance.

Most wax polishes are a mixture of natural (organic) waxes (e.g. beeswax, whale oil and carnauba wax) and synthetic waxes (e.g. from petroleum products).

All natural waxes contain SILICON to a greater or lesser extent. Some products however are described as Silicon polishes, but are completely inorganic. For example:

ORGANIC BASED WAX ....

INORGANIC BASED WAX ....

SILICON-FREE POLISH.   It is *very desirable* that only this type of polish is used in the body shop. The reason for this is:

....

....

....

Name ONE type of silicon-free polish:

....

Steel wheels

If badly marked or rusted: ....

....

Aluminium wheels

It is essential that all appropriate safety precautions are observed since aluminium cleaners tend to be very

....

Tyres

To cover overspray use ....

To enhance appearance (no overspray) ....

Chromium plate

Some standard wax based polishes are suitable for cleaning and protecting chromium plating.

Heavy oxidisation may be removed by gentle rubbing with ....

....

Plastic parts (e.g. bumpers)

Plastic is widely used for body sections on many cars. Over time they can appear faded and may show white chalky marks from the overspill of old polish. How might their appearance be restored?

....

Typical product ....

Show on the drawing which body panels are likely to be made of plastic and may need restoration by trim gel.

USED CAR INTERIOR VALET

Mention some of the major points that need to be considered when dealing with an older vehicle that has seen hard use.

Safety

As always, proper attention should be given to safety and product instructions.

CARPETS ....

....

Note: For shampoo cleaning, treat as heavily soiled seat fabrics.

List FOUR other areas to be vacuumed:

....

RUBBER MATS AND RUBBER CARPET INSERTS

Clean using ....

Improve appearance by ....

MUD, DOG HAIRS, etc.

Remove by ....

HEADLINING

These are mostly made from fabric (usually nylon) or vinyl. Some cleaning products are suitable for only one or the other, while some will suit either. It can therefore be sensible to test the product before use. Test by:

....

....

List three different types of cleaner and how to apply them:

Fabric ....

Vinyl ....

Universal ....

Aerosols

The use of aerosol type sprays is quite common, being convenient and economical for the relatively small area coverage inside cars. Their use, though, can present particular problems, TWO examples are:

1. ....

2. ....

Suggest THREE ways to minimise such hazards:

1. ....

....

2. ....

....

3. ....

....

Interior plastic and vinyl (hard) surfaces

List THREE typical 'hard' surfaces:

1. ....

2. ....

3. ....

Many of the products made for cleaning and renovating this type of surface have a high silicone content. It is recommended that such products are NOT used:

1. In the vicinity of the body shop ....

....

2. On steering wheel, pedals and mats etc. ....

Name ONE non-silicone plastic and vinyl cleaner and say how it may be applied:

....

....

Seats

Seats (and door trim panels) are usually one of three types:

1. .... 2. .... 3. ....

Fabric Upholstery

Lightly soiled fabric can be cleaned by hand, using a cleaner as for fabric headlining. Heavily soiled fabrics are likely to respond better to the use of a spray/extraction machine. This is a hot water/detergent spray combined with a vacuum extractor to remove the moisture and the dirt. What is indicated by the three types of arrow shown below?

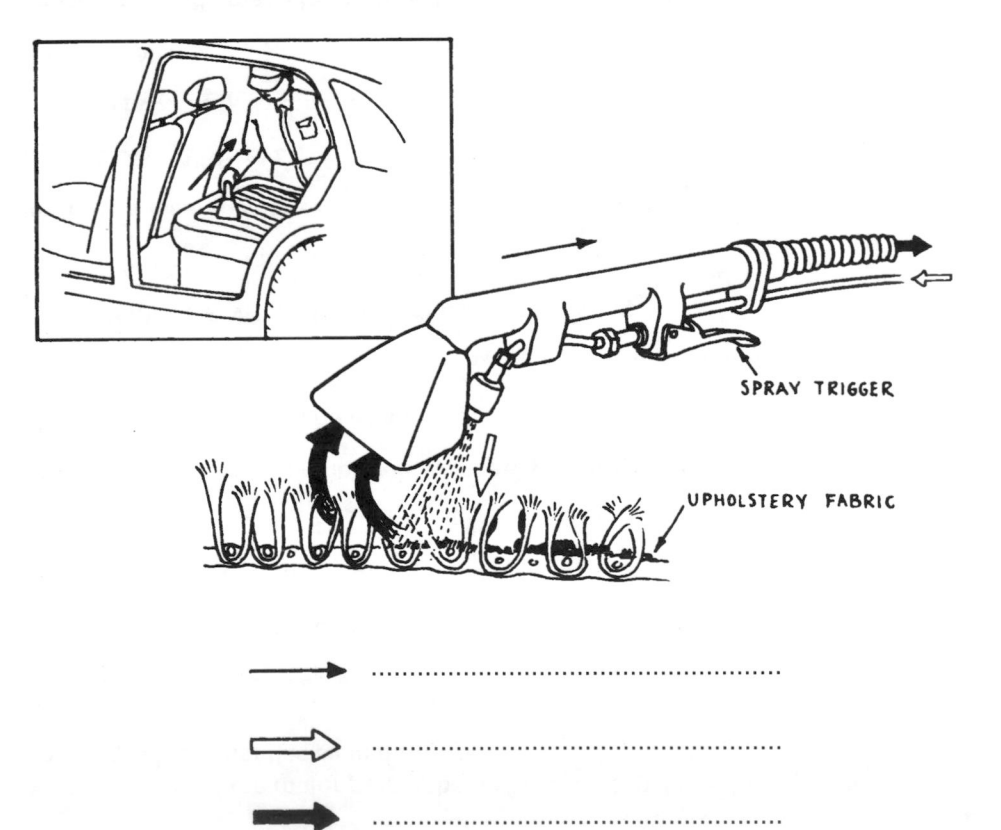

SPRAY TRIGGER

UPHOLSTERY FABRIC

→ ....

⇨ ....

➡ ....

Heavily Soiled Fabric

Typical cleaning sequence would be:

1. ....
....

2. ....
....

3. ....

Seats must not become too wet during cleaning, because:

1. ....

2. ....

VINYL UPHOLSTERY

Typical cleaning sequences would be:

1. ....

2. ....

3. ....

LEATHER UPHOLSTERY

IMPORTANT: Leather must NOT be treated with petroleum-based solvents, detergents or caustic soaps. Otherwise permanent (and expensive!) damage can result.

Leather may be cleaned by washing using warm water (not too much), and special soap or cleaner.

SOAP ....

CLEANER ....

WASH ....

REMOVE RESIDUE ....

DRY ....

Glass

Traditionally the way to clean glass has been by means of water and wash-leather. However, largely as a result of pollutants in the atmosphere, the process is now rarely effective. For example, the insides of windscreens can become particularly badly smeared as a result of:

1. ....

2. ....

3. ....

Suggest ONE product suitable for cleaning smeared glass:

....

Glass Cleaning Procedure

1. ....

....

2. ....

....

3. ....

4. ....

....

5. ....

6. ....

Some operators use a very inexpensive alternative to cloth for polishing glass:

....

Door Seals

The rubber seals around the edges of doors can sometimes look tired and faded. Their appearance may be improved by:

....

....

Polishing Cloth

Most widely used is ....

There are, of course, varying grades of mutton cloth.

QUALITY CHECK

Before a car is released to a customer it is essential to make a careful visual inspection. This is to ensure that the customer's instructions have been fully complied with and that the work complies with the firm's own quality standards.

List TWELVE of the items that would need to be checked in the case of a full valet:

1. *Body must be free of all dirt, wax, or grease.*

2. ....

3. ....

4. ....

5. ....

6. ....

7. ....

8. ....

9. ....

10. ....

11. ....

12. ....

If the valeter is in doubt about the quality of any area of finished work, what should be done?

....

....

....

CAR VALET RECORD – USED CAR EXTERIOR AND ENGINE COMPARTMENT   JOB No. ....

Make ....   Model ....   Year ....

Paint colour ....   type ....   hard/soft ....

Area	Method	Product	Dilution	PPE	Safety
Body wash					
Underwings					
Wheels					
Body clean					
T-cut					
Body wax					
Chromium					
Glass					
Plastic					
Tyres					
Engine compartment					

Job satisfactorily completed to all required standards.

Any waste or surplus materials have been disposed of properly and safely.

Signature ....
             (Student)   (Assessor)

CAR VALET RECORD – USED CAR INTERIOR

JOB No. ....

Make ....

Model ....

Year ....

Seat material ....

Headlining ....

Area	Method	Product	Dilution	PPE	Safety
Vacuum interior					
Seats					
Carpets					
Rubber mats					
Headlining					
Instrument panel					
Fascia					
Door trim					
Boot					

Job satisfactorily completed to all required standards.

Any waste or surplus materials have been disposed of properly and safely.

Signature ....

(Student)   (Assessor)

PRESSURE CLEANERS

High-pressure cleaners are very versatile valeting tools.
Compared with hand washing they are:

....

....

There are TWO types of pressure cleaners:

....

Both types use a pump to increase water delivery pressure above that of the
mains water supply. (The hot water type can also deliver cold water.) Output
pressure can be adjusted and in addition they are also able to 'bleed-in' to the
output:

....

....

Why does the motor trade generally prefer the HOT water pressure cleaner to the
cold water type?

....

....

List SIX important points when using a pressure cleaner to apply a chemical mix
to a car:

1. ....

2. ....

3. ....

4. ....

5. ....

6. ....

List THREE important points when rinsing cleaner from car:

....

Label the major parts of the pressure washer shown below:

Inspect a power washer and determine the following information:

Make ....   Model ....

Max. output temp. ....   Max. output pressure ....

Fuel used ....   Reservoir capacity ....

Date of most recent check by qualified electrician ....

Safe Working Using Hot Water/Steam Pressure Cleaners

There are THREE main sources of potential danger when using these machines:

1. ....

2. ....

3. ....

What precautions help to guard against accidents when using hot water/steam pressure cleaners?

MAINS ELECTRICITY

1. Supply must be via ....

2. Cable and plug condition ....

3. Cable routing ....

4. Earth-bonding of spray lance and metal shield ....

....

Note: Electrical risks can be minimised by using a fixed power wash unit and probably a longer hose to lance; or by using 110 volt supply.

HIGH PRESSURE

1. To minimise recoil from lance pressure ....

....

2. Do not direct spray at ....

3. Do not direct high-pressure fine-jet spray on ....

4. Lance trigger ....

5. High-pressure hose disconnection ....

....

HIGH TEMPERATURE

1. Pressure hose disconnection ....

2. Avoid spraying ....

3. Spray temperatures ....

....

AUTOMATIC CAR WASH

Many car owners make frequent use of automatic car washers to keep their cars clean. Some repair shops also use them for car cleaning before it is returned to the customer.

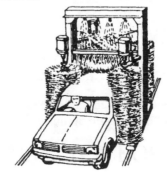

FOUR reasons for this are:

....

....

....

....

Describe TWO types of automatic car washers.

WASHER TYPES

Brushless ....

....

Revolving nylon brushes ....

....

RANGE OF SERVICES

The range of services offered by automatic car wash operators varies considerably – usually according to cost.

Examine the various services available at a local automatic car wash and list them below.

Type of washer ....

Vehicle preparation by operator

....

....

....

Automatic machine operations

....
....
....
....
....
....
....

Cost of full treatment .... Time taken ....

Some filling stations and automatic car wash sites offer other cleaning facilities, primarily for customer use. Very often they are coin, or token, operated and may include:

....
....
....
....
....

Curious as it may seem, the brushes of automatic car washers need quite frequent cleaning. Why is this and how is it done?

....
....
....?..

List THREE other items of maintenance needed on a daily basis:

....
....
....

HAND WASHING

The main benefits of washing vehicles by hand is that it is often very convenient, especially for the occasional job, and the facilities and equipment need only be minimal.

The TWO systems commonly used in the motor trade are:

1. Bucket and sponge ....

....
....

2. Water-fed 'Flexybrush' ....

....
....
....
....

With both systems, best results are obtained by using plenty of water and a gentle sweeping action.

Apply soap-and-water mix starting at the ....

Rinse with clean water, starting at the ....

SOLVENT PROPORTIONS

Why is it important in any valeting operation to use only the recommended proportion of solvents or soaps, even if say the vehicle is particularly dirty?

....
....
....
....

Chapter 6

Routine Maintenance and Inspection

SCHEDULED SERVICES AND INSPECTIONS

New vehicles must be in first class running order when delivered to the customer. They then must be regularly serviced to maintain this condition.

Name FOUR typical services scheduled by a manufacturer:

1. ....

2. ....

3. ....

4. ....

Suggest typical mileage service intervals:

....

....

What would be typical time-based service intervals?

....

....

Garage servicing does not mean that the owner has no need to carry out weekly or two-weekly checks.

List typical owner weekly checks that should be carried out by the driver:

....

....

....

State the PURPOSE of the FOUR manufacturer-recommended service/inspection schedules.

1. Pre-delivery inspection (p.d.i.)

....

....

....

....

2. First service

   This is usually scheduled at 500 or 600 miles, or even 1000 miles. (Some manufacturers incorporate this service with the pre-delivery inspection.)

....

....

....

3. Mileage-based service

....

....

....

....

4. Time-based service

....

....

....

The modern trend is to extend substantially the mileage service intervals.

List FOUR items that contribute towards this trend:

....

....

....

....

ITEMS AND SYSTEMS REQUIRING SERVICE CHECKS

Identify the vehicle systems shown below that would require basic service checks.

The illustrations are not of any one specific type of car.

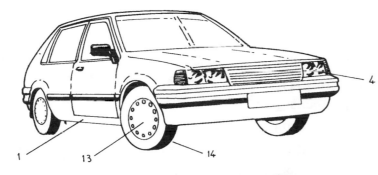

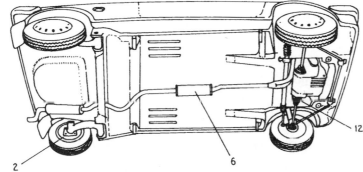

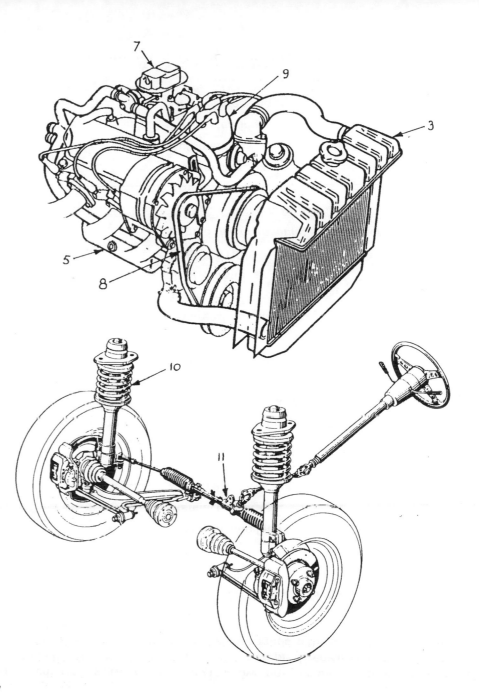

1.	8. ....
2.	9. ....
3.	10. ....
4.	11. ....
5.	12. ....
6.	13. ....
7.	14. ....

INTERPRETATION OF VEHICLE SERVICING DATA

Before or during the servicing of a vehicle it is always necessary to know some technical data about the vehicle. Using a workshop manual, complete the data asked for on this page for a typical vehicle requiring service. If you do not understand some of the technical terms given, ask your colleagues in the class/workshop (i.e. *Maintain effective working relationships*).

VEHICLE MAKE ....

MODEL ....

GENERAL DATA

ENGINE

Cylinder arrangement ....

Bore ....

Stroke ....

Compression ratio ....

Naturally aspirated/Turbocharged

Oil filter   *type* ....

Air filter   *type* ....

Fuel filter   *type* ....

CLUTCH

Type ....

Adjustment ....

SUSPENSION

Front wheel alignment ....

TYRES

Type ....

Pressures cold   *Front* ....

                    Rear ....

ELECTRICAL

Battery   *type* ....

Cold crank ....

Reserve capacity ....

Alternator   *type* ....

Output ....

Starter motor   *type* ....

Power ....

TECHNICAL DATA

ENGINE

Type/Capacity ....

Firing order ....

Idle speed ....

Idle CO content ....

IGNITION

Ignition timing at idle ....

Vacuum   *connected/disconnected*

Coil type ....

Primary resistance ....

Spark plugs, type ....

Gap setting ....

Tightening torque ....

COOLING SYSTEM

Pressure cap pressure ....

Thermostat temperature ....

ELECTRONIC FUEL INJECTION

Type ....

....

Fuel line pressure ....

Throttle potentiometer voltage

    Throttle closed ....

    Throttle operated ....

FUEL

Grade/Type ....

CAPACITIES, FLUIDS and LUBRICANTS

CAPACITIES   Litres

Fuel tank ....

Engine oil and filter change ....

Manual gearbox ....

Automatic   *Refill* ....

                Dry ....

Cooling system ....

Windscreen washer reservoir ....

Power steering reservoir ....

FLUIDS

Brake/Clutch fluid   *Top up*

....

Antifreeze solutions

Type ....

Concentrations ....

LUBRICATION   Type

Engine *Use oil meeting specification standards of*

....

Viscosity ....

Manual gearbox ....

Automatic ....

Power steering ....

General greasing ....

....

VEHICLE SERVICE INSPECTION DATA

Use the data on this page or similar data to check the condition of various vehicles

PRE-DELIVERY INSPECTION ✓

Check heater and cooling system, oil cooler and all fuel pipes, tank cap, hoses and connections for leaks	
Check engine coolant level and antifreeze strength	
Check drive belts for deflection	
Check engine idle speed, CO emission and choke operation	
Check battery electrolyte levels and terminals for security	
Check engine, transmission and differential oil levels and power steering fluid reservoir level, top up as necessary	
Check brake and clutch fluid levels	
Check brake and clutch system pipes, hoses and connections for leakage	
Check brake and clutch pedal free play and parking brake travel	
Check exhaust system and mountings for security and leakage	
Check all steering and suspension connections for security	
Check all electrical systems for correct operation	
Check all lights for damage, correct operation and alignment	
Check operation of window regulators, bonnet and boot lid	
Check all locks and doors for correct closure and lubricate as required including fuel filler door	
Check windscreen wipers/washers for correct operation/direction and top up washer fluid level	
Check tyre pressures and torque setting of road wheel nuts	
Lubricate propeller shaft grease points (as applicable)	
Check tool kit	
De-wax, clean and polish vehicle internally and externally	
Check operation of seat belts, condition of webbing and security of anchorage points	
Fit carpet as necessary	
Ensure interior is free from grease and dirt	
Remove brake disc anti-corrosion covers	
Check for defects on paintwork and report	
Road test and report	

'CHECK' means check and adjust, top up or report.

1 MONTH/600 MILE SERVICE ✓

Check valve clearances (as applicable)	
Check drive belts for deflection	
Check engine coolant level and strength	
Inspect CB points and rotor (as applicable)	
Check ignition timing, dwell angle and advance operation	
Check engine idle speed, CO emission and choke operation	
Inspect fuel tank cap, fuel lines and connections	
Check throttle positioner/dash pot (throttle damper setting/speed)	
Check engine, transmission and differential oil levels and power steering fluid reservoir level, top up as necessary	
Check brake and clutch fluid levels	
Check brake and clutch pedal free play and parking brake travel	
Check brake and clutch system pipes, hoses and connections for leakage	
Check heater and cooling system, oil cooler, and all fuel pipes, tank cap hoses and connections for leakage	
Check all electrical systems for correct operation	
Check all lights for damage and correct operation and alignment	
Check windscreen wipers/washers for correct operation/direction and top up washer fluid level	
Tighten bolts on chassis and body	
Check tyre condition, pressures and torque setting of road wheel nuts	
Check operation of seat belts, condition of webbing and security of anchorage points	
Check bodywork for abrasions or damage which if left unattended may lead to corrosion	
Road test and report	

'CHECK' means check and adjust, top up or report.

© TOYOTA

159

VEHICLE SERVICE INSPECTION DATA – MAJOR SERVICES

The schedule shown below covers most service checks required on a family car at mileage periods of 6000 miles up to 36 000 miles.
In the three columns tick the items that require checking at 6000 miles, 18 000 miles and 36 000 miles.

ELEVATED

Item		MILES × 1000 6	18	36
Engine: oil	drain			
Engine: oil filter	renew			
Gearbox: oil	check/top up			
Rear axle: oil	check/top up			
Steering linkage: condition	check/report			
Steering rack: mountings	check/tighten			
Steering rack seals: condition	check/report			
Steering system: leaks	check/report			
Steering linkage/mountings: security	check/report			
Suspension linkage/mountings: security	check/report			
Front wheel bearings	check/repack/adjust			
Shock absorbers: condition/leaks	check/report			
Suspension system: safety check	check/report			
Exhaust system: condition	check			
Parking brake linkage: grease	lubricate			
Parking brake linkage	adjust			
Brake hydraulic system: safety check	check/report			
Clutch: free play	check/adjust			
Brake hydraulic system: fluid	drain/refil			
Brake hydraulic system: components	renew			
Disc brakes: pads/discs condition	check/renew			
Drum brakes: linings/drums	check/renew			
Drum brakes	check/adjust			
Wheel nuts/bolts	check/tighten			
Tyres: condition/tread depth	check/report			
Tyres: pressures	check/adjust			

SHELL

ON FLOOR

Item		MILES × 1000 6	18	36
Engine: oil	refill			
Engine: oil	check/top up			
Throttle linkage: oil can	lubricate			
Throttle pedal: oil can	lubricate			
Parking brake: travel	check/adjust			
Brake reservoir: fluid	check/top up			
Drive belt: condition	check/report			
Drive belt: tension	check/adjust			
Timing belt: tension	check/adjust			
Valve clearances	check/adjust			
Spark plugs: gap	check/adjust			
Spark plugs	renew			
Distributor: contact breaker	check/renew			
Distributor: shaft	lubricate			
Distributor: advance/retard mechanism	check/report			
Distributor cap/HT leads/coil	clean			
HT leads	check/secure			
Oil filler: breather cap	clean/renew			
Crankcase ventilation valve	clean/test			
Air cleaner: element(s)	renew			
Inlet/exhaust manifold: security	check/tighten			
Radiator/heater hoses: condition/leaks	check/report			
Cooling system	check/top up			
Anti-freeze: content	check/report			
Battery terminals	clean/smear			
Battery electrolyte	check/top up			
Battery specific gravity	check/report			
Locks/catches/hinges: oil can	lubricate			
Locks/catches/hinges: operation	check/report			
Front wheel alignment	check/adjust			

PRE-MOT TESTS

GENERAL LEGISLATIVE REQUIREMENTS FOR ROAD VEHICLES

ITEMS TESTED

The items that are tested during an MOT test are listed below and on the next page.

On a pre-MOT test these items would be checked and any faults rectified before the vehicle was submitted for the test.

State for each item the expected legislative requirement (read the MOT tester's manual using the references given below) and then carry out a pre-MOT test on a vehicle using the sheet on the next page.

Item	Manuel ref. Cars & LGV	
LIGHTING EQUIPMENT		
Front & rear lamps etc.	1.1	....
Headlamps	1.2	....
Headlamp aim	1.6	....
Stop lamps	1.3	....
Rear reflectors	1.4	....
Direction indicators & hazard lamps	1.5	....
STEERING AND SUSPENSION		
Steering control	2.1	....
Steering mechanism/system	2.2	....
Power steering	2.3	....
Transmission shafts	2.5	....
Wheel bearings	2.5	....
Front suspension	2.4, 5	....
Rear suspension	2.4, 6	....
Shock absorbers	2.7	....
BRAKES		
ABS warning system/control	3.4	....
Condition of service brake system	3.3, 5, 6	....

Item	Manual ref. Cars & LGV	
Condition of parking brake system	3.1, 2, 5	....
Service brake performance	3.7	....
Parking brake performance	3.7	....
TYRES & WHEELS		
Tyre type	4.1	....
Tyre load/speed ratings (Class V & VII)	4.1	....
Tyre condition	4.1	....
Roadwheels	4.2	....
SEATBELTS		
Mountings/Condition/Operation	5.1	....
GENERAL		
Driver's view of the road	6.1	....
Horn	6.2	....
Exhaust system	6.3	....
Exhaust emissions	6.4	....
General vehicle condition	6.5	....
Mirrors	6.6	....
Fuel system	6.7	....
Registration plates and VIN numbers	6.8	....

CLASS IV M.O.T INSPECTION SHEET

Use this form (while taking the recommended route around the vehicle) to indicate the condition of each item - **Pass** or **Fail** - and for Advising customers of components (whether testable or not) which are already or may become defective in the near future. THE RELEVANT VT20 or VT30 **MUST BE ISSUED TO THE CUSTOMER. VT 20/ 30 No.**

Make:	Model:		Year:	Reg No.::		Nominated Tester:
Rec Mileage:	Colour:		Weight:	Petrol/Diesel?	CC:	
VIN/Chassis No:				VTS No:		
Is the vehicle of right type and condition to proceed?		Diesel vehicle- Customer questionaire?		Is assistant and all necessary equipment available?		Date:
						V5 Checked?

Pass	Fail	Advise	Testable item		Comments
			Rear view mirrors	1	
			Driver's seat and back rest	2	
			Note recorded miles of speedo	3	
			Front pass seat and backrest	4	
			Both doors/internal security	5	
			Steering wheel/ mechanism/ cond	6	
			Power steering (if fitted) *	7	
			Servo operation (if fitted) *	8	
			Footbrake operation/ condition	9	
			ABS system warning light (if fitted)	10	
			Handbrake operation/ condition	11	
			Light switches/ warning lights	12	
			Screen wipers/ washers operation	13	
			Wind screen condition / vision	14	
			Horn control/ operation	15	
			Belt requirements front/ rear	16	
			Belt cond/ oper/ anchorages	17	
			No plate condition/ spacing	18	
			Side - H/ lamp type/ cond/ operation	19	
			Indicators cond/ operation/ rate	20	
			Hazard warning operation/ rate	21	
			Headlamp cond/ aim. (use equip)	22	
			O/S/F s/absorber & body damage	23	
			O/S/F tyre wall/ size/ valve/ fit	24	
			Ball joint/ wheel security/ cond	25	
			O/S wiper blade cond/screen cond	26	
			O/S repeater lamp cond (if fitted)	27	
			O/S/F door security/ext. mirror	28	
			O/S/F out-inn sills/ floor/ belt mntgs	29	
			O/S/R door security/rear back rest	30	
			O/S/R out-inn sills/ floor/ belt mntgs	31	
			O/S/R tyre wall/ size/ valve/ fit	32	
			O/S/R wheel security/ condition	33	
			O/S fuel cap	34	
			O/S/R shock absorber	35	
			Rear tail gate/ boot/ door security	36	
			R floor/ belt mtgs/ shock abs mntgs	37	
			Tail lights cond/ operation	38	
			Stop/ Fog cond/ operation	39	
			Indicators cond/ operation/ rate	40	
			Hazard warning operation/ rate	41	
			No plate lights and reflectors	42	
			No plate condition/spacing	43	
			N/S/R s/absorber & body damage	44	
			N/S fuel cap	45	
			N/S/R tyre wall/ size/ valve/ fit	46	
			N/S/R wheel security/ condition	47	
			N/S/R out-inn sills/ floor/ belt mntgs	48	
			N/S/R door security/ floor/ belt mntgs	49	
			N/S/F door security/ ext. mirror	50	
			N/S/F out-inn sills/ floor/ belt mntgs	51	
			N/S repeater lamp cond (if fitted)	52	
			N/S/F tyre wall/ size/ valve/ fit	53	
			Ball joint/ wheel security/ cond	54	
			N/S wiper blade cond/screen cond	55	
			N/S/F shock absorber	56	
			Under bonnet brake system/ mntgs		

Pass	Fail	Advise		Testable item
			57	Pipes/ hoses/ master cylinder/ servo *
			58	Handbrake mechanism
			59	Suspension mounts/ cond/ corrosion
				Battery security/vin No
			60	Fuel system leakage/ security *
			61	Exhaust system leaks *
			62	Servo vacuum hose (if fitted) *
			63	PAS pump-drive-pipes (if fitted) *
			64	Diesel pump-pipes-cambelt-oil level
			65	Steering components (rock str wheel)
			66	Repeat and check from under vehicle
			67	Str box rack cond/ wear/ mountings
			68	Ball joints cond/ wear/ play
			69	Track rods-steer arm cond/ security
		△	70	O/S/F tyre wall/ fit/ tread
		△	71	O/S/F wheel condition
		△	72	C/S/F wheel bearing condition
		△	73	O/S/F drive shaft / CV joint
		△	74	O/S/F suspension cond/ security
		△	75	N/S/F tyre wall/ fit/ tread
		△	76	N/S/F wheel condition
		△	77	N/S/F wheel bearing condition
		△	78	N/S/F drive shaft/ CV joint
		△	79	N/S/F suspension cond/ security
			80	Lock to lock checks (turn plates)
		△	81	Suspension check - shake O/S wheel
		△	82	Suspension check - shake N/S wheel
		△	83	O/S/R tyre wall/ fit/ tread
		△	84	O/S/R wheel condition
		△	85	O/S/R wheel bearing condition
		△	86	O/S/R drive shaft/ CV joint
		△	87	O/S/R suspension cond/ security
		△	88	N/S/R tyre wall/ fit/ tread
		△	89	N/S/R wheel condition
		△	90	N/S/R wheel bearing condition
		△	91	N/S/R drive shaft/ CV joint
		△	92	N/S/R suspension cond/ security
		*#	93	O/S/F brake components cond/leak *#
		*	94	Front chassis members/ suspension
		*#	95	N/S/F brake components cond/leak *#
		*	96	Mid chassis members/ suspension
		*#	97	O/S structure/ pipes/ cables/ hoses *#
		*#	98	O/S/R brake components/ cond/leak *#
		*	99	R/chass members/ susp/ tank/carrier
		*#	100	N/S/R brake components/ cond/leak *#
		*#	101	N/S structure/ pipes/ cables/ hoses *#
		*	102	Exhaust condition/ leaks/prop shaft
		*	103	H/brake mechanism/ linkage/ cables
		*	104	O/S/F footbrake performance check *
		*	105	N/S/F footbrake performance check *
		*	106	Both front brakes balance check *
		*	107	Handbrake O/S performance check *
		*	108	Handbrake N/S performance check *
		*	109	O/S/R footbrake performance check *
		*	110	N/S/R foot brake performance check *
		*	111	Both rear brakes comparison check *
		*	112	Exhaust emission (last or first) *

Readings Obtained:

WARNING In my opinion the vehicle is dangerous to drive because of the following defects:-

* Engine running   # Footbrake applied   △ Jacking beam in use

E. + O.E.   Copyright 1/95 M O T Management Service Yew Tree Cottage Acton Nantwich CW5 8LG Tel: 01270 629686 Fax 01270 628881

VEHICLE BODY INSPECTION

When carrying out an inspection service on the vehicle body, faults could be categorised under the headings below.

List common faults and where they might be expected to occur:

1. Paintwork

....
....
....
....
....
....

2. Corrosion to body panels

....
....
....
....
....

3. Underbody/chassis members for damage or corrosion

....
....
....
....
....

4. Door cover lock latch and hinge operation

....
....
....
....
....
....

5. Body and cab securing and locking devices

....
....
....
....

6. Seat belts, condition and security

....
....
....
....
....

What lubrication or adjustments might reasonably be expected to be done during a vehicle body check and service?

....
....
....
....

163

VEHICLE BODY INSPECTION

Vehicle make .... Model ....

Vehicle Exterior

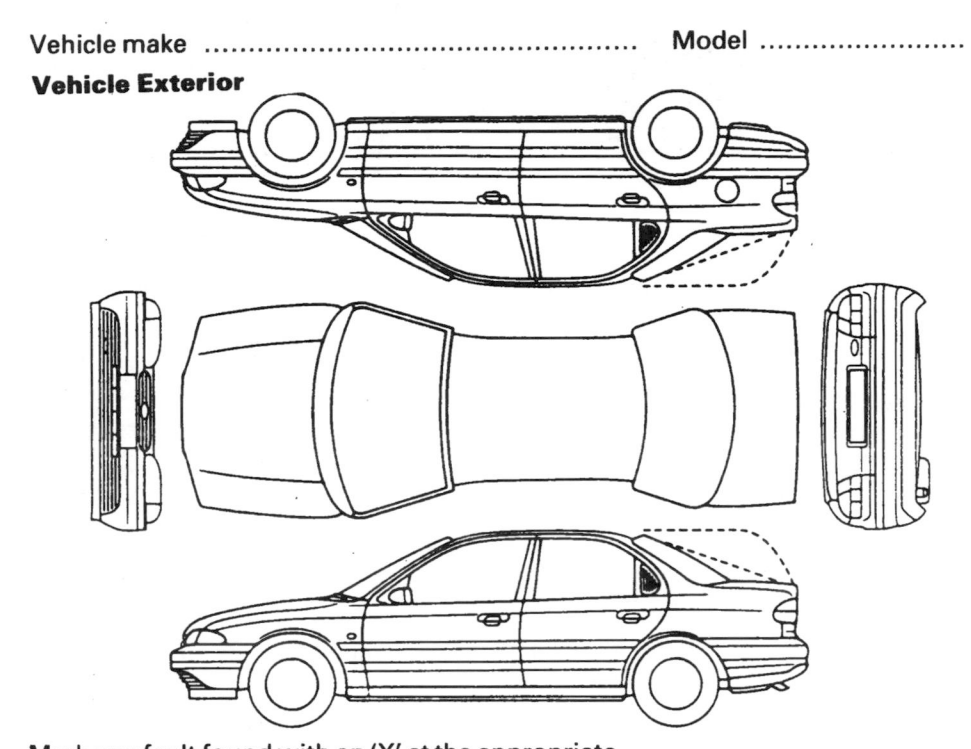

Examine a vehicle in reasonable bodily condition and complete the table below.
Note: This exercise is related to routine maintenance, not body repairing.

Vehicle Interior

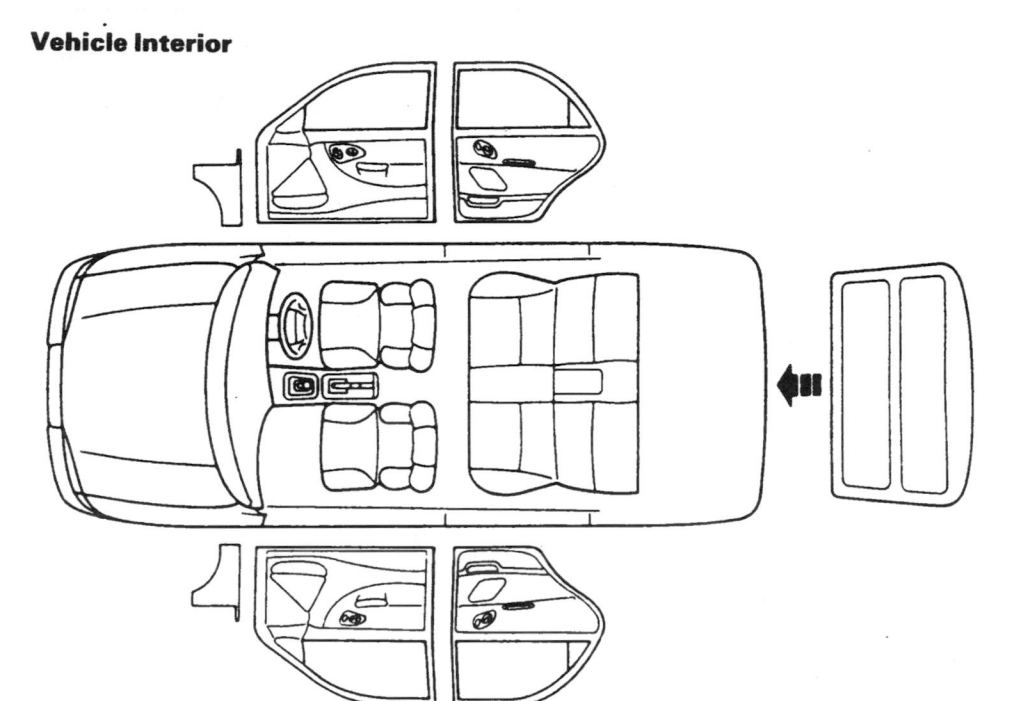

Mark any fault found with an 'X' at the appropriate position on the diagrams and annotate a consecutive number. Enter fault and rectification details by this consecutive number in the box opposite.

DAMAGE/CONDITION CHECK

Check exterior for damage/condition of paint, sheet metal and trim ☐

Check interior for damage/soilage of trim and upholstery ☐

Check storage compartments for correct fit and closure of lids ☐

Check carpets (including luggage compartment) for correct fit and security ☐

	Causal Part	Defect	Method of Rectification	Corrected
1				
2				
3				
4				
5				
6				
7				
8				
9				
10				

ENGINE SERVICING

Oil Level Checks

The engine oil level must be frequently checked. It is therefore essential that the dip stick level readings are understood and the engine oil capacity known.

Indicate on the bottom of the dip stick the usual oil level markings and what they represent.

State typical engine oil capacity.

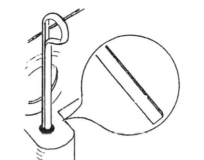

Vehicle make	Model	Engine size (cc)	Oil capacity (litres)

Breathers and Cap Filters

On routine maintenance inspections, a check commonly overlooked is the cleaning of breathers, pipes and their filters.

How should the filler cap shown be serviced?

....

....

....

Why should the cap be cleaned?

....

....

....

....

Name and indicate the parts that require cleaning. Note that broad arrows show air flow.

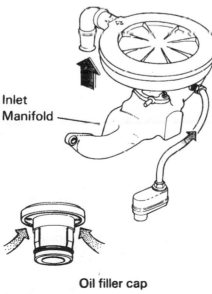

Inlet
Manifold

Oil filler cap

Oil and Filter Change

Oil filter fitments may be basically of two different types. Name those shown below:

1. ....   2. ....

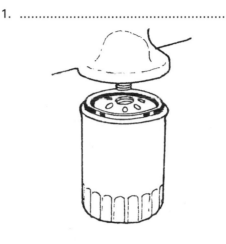 

How do these two types of filter differ?

1. ....   2. ....

....

....

....

How does fitting with regard to the sealing ring differ?

1. ....   2. ....

....

....

How are the units tightened?

1. ....   2. ....

....

....

ENGINE INSPECTION

Oil Leaks

Indicated on the lubrication diagram of the engine below are typical points where oil leaks may occur.

Examine the sketch and name the possible leakage points that have been numbered.

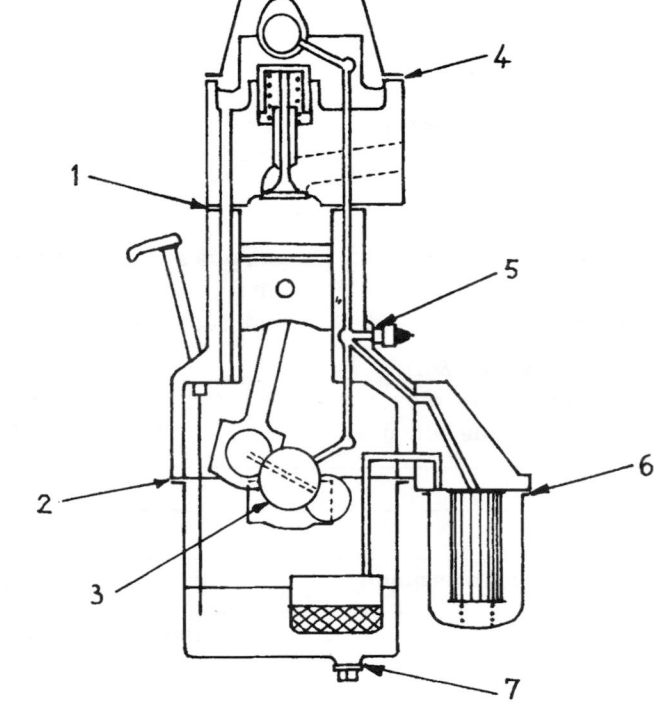

1. ....
2. ....
3. ....
4. ....
5. ....
6. ....
7. ....

Timing Belt Tension

Most modern engines are fitted with internally notched timing belts. These should be periodically checked for correct tension and condition.

How should a timing belt be checked for correct tension?

....
....
....
....
....
....
....
....

What faults would require the belt to be changed?

....
....
....
....

At what mileage should a timing belt typically be changed?

....

Name the parts indicated on the drive belt assembly

1. ....
2. ....
3. ....
4. ....
5. ....

What may occur to the engine if the timing belt is not changed at the recommended mileage?

....
....
....

Valve Clearance

The valves in an engine require clearance, there are two reasons for this:

1. ....

2. ....

Obtain data for a conventional four-cylinder engine for which the valve clearances will need to be adjusted during servicing.

Make of engine .... Type ....

State the engine's valve clearances in mm and inches.

Inlet valve   mm .... in. ....

Exhaust valve   mm .... in. ....

When the following valves are fully open	8	7	6	5	4	3	2	1
Check the clearance of valve nos.	1	2	3	4	5	6	7	8

The above method uses a 'law of 9' and is suitable for most four-cylinder engines.

....

....

....

....

Why do some engines not require the valve clearances to be adjusted?

....

....

....

....

Methods of valve adjustment

Below are shown some typical engine valve operating mechanisms. In each case show the position of the feeler gauge and explain how adjustment is made.

(a)

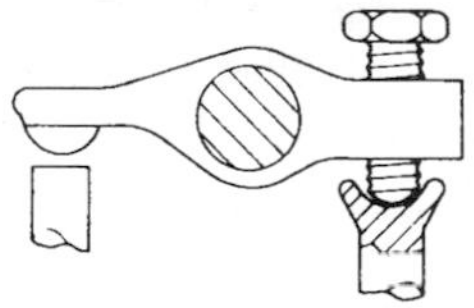

(b)

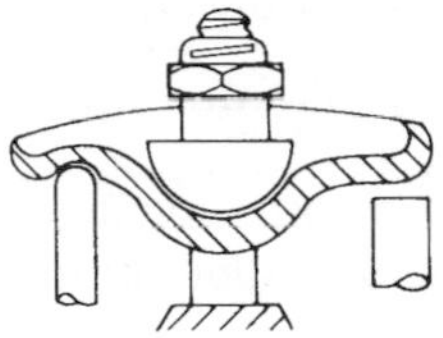

(c)

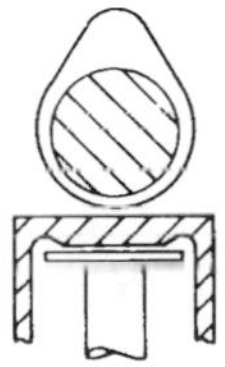

(d)

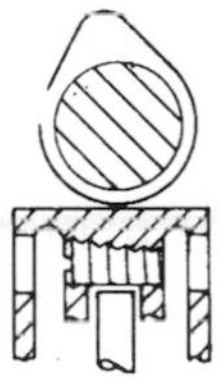

(e)

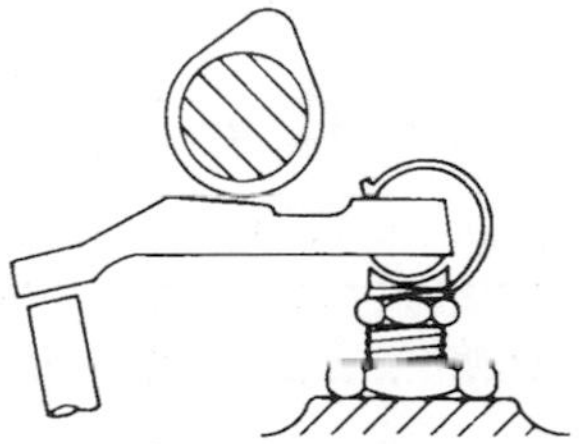

COOLING SYSTEM MAINTENANCE

Servicing the cooling system may largely consist of a visual check, with possibly a change of coolant or component.

The coolant level is usually checked by observing the fluid level in the translucent expansion bottle. What is the usual correct level of fluid in this bottle?

....

....

List TEN items that should be checked for leakage:

1. ....
2. ....
3. ....
4. ....
5. ....

6. ....
7. ....
8. ....
9. ....
10. ....

Note the direction of water flow through the system and name the parts indicated:

Hot

Cool

© ROVER

What effect does ageing have on the hoses?

....

....

When should a cooling system be drained and flushed?

....

....

How is the proportion of anti-freeze to water content usually checked?

....

....

How should anti-freeze/inhibitor be prepared before being put into the system?

....

....

....

....

Air locks occur in many coolant systems when refilled. How generally are these de-aerated? Consider the case of a car having air bleed valves.

....

....

....

....

The radiator matrix may eventually become clogged with dirt and flies. How should these be removed?

....

....

IGNITION SYSTEM MAINTENANCE

Name the main parts of the simple coil-ignition system shown and draw the primary and secondary circuits leads.

What is the firing order of the rotor turns anti-clockwise? ....

What items should be visually checked on the ignition system before dismantling and after removal of the distributor cap?

....

....

....

....

What is the speed relationship between the distributor and the crankshaft?

....

Spark Plug

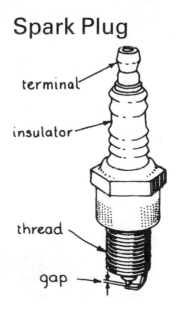

terminal

insulator

thread

gap

Distributor

If replacing an ignition module why should you smear special silicone grease between the module and the mounting surface?

....

....

....

....

....

What is the function of the ignition module?

....

At what typical mileage should the spark plugs be removed and checked? ....
What basic service should be given to a plug which is removed for cleaning?

....

....

....

....

....

....

Name the parts indicated

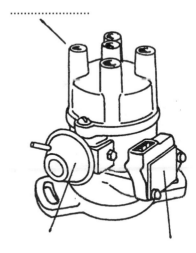

AIR SUPPLY MAINTENANCE

The engine air cleaner has two functions:

1. *To remove foreign matter (dust, grit) from the air before it enters the* ....

   carburettor and engine. ....

2. ....

Below are shown two completely different types of air cleaners. Name the types and describe how each should be serviced:

1. ....

   ....

   ....

   ....

   ....

2. ....

   ....

   ....

   ....

   ....

   ....

1. ....

Indicate the air flow through the filter.

Oil level do not overfill

Wire mesh

2. ....

The usual mileage between services is ....

Air Temperature Control

In European countries modern air cleaners have some means of allowing warm air to enter the engine during the winter period. This control may be a flap valve (operated manually) or by a thermostat.

Manually controlled air cleaner

What service checks should be made?

....

....

....

Temperature controlled air cleaner

The thermostat moves the valve deflector plate to a position which will allow a mixture of hot/cold air to be supplied at a temperature between 23° and 28°C.

What service checks should be made?

....

....

....

....

FUEL SYSTEM – PETROL

Show on the drawing below, a typical petrol fuel system. Name the parts and indicate where service attention is likely to be required.

What checks should be carried out to the fuel system during a routine service?

....
....
....
....
....
....
....
....
....

The diagrams show a typical constant depression carburettor. State the checks associated with the numbered parts.

FUEL INJECTION

A fuel injection layout is shown below. At this stage it is only necessary to check if all parts are secure and fitting correctly.

Name the parts indicated.

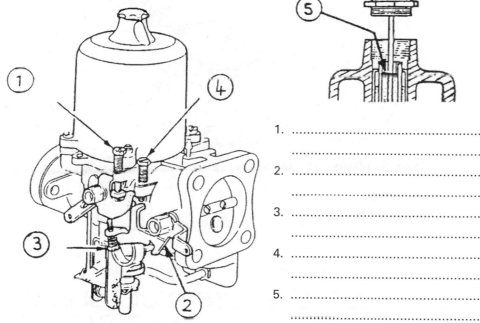

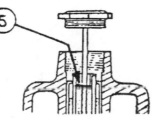

1. ....

2. ....

3. ....

4. ....

5. ....
....

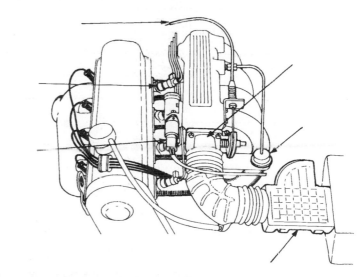

NOTE: If the fuel injection system is to be worked on, the fuel supply system must first be depressurised.

FUEL SYSTEM –
COMPRESSION IGNITION (DIESEL)

An in-line pump diesel fuel system is shown. Note the direction of fuel flow around the system. Name the parts indicated.

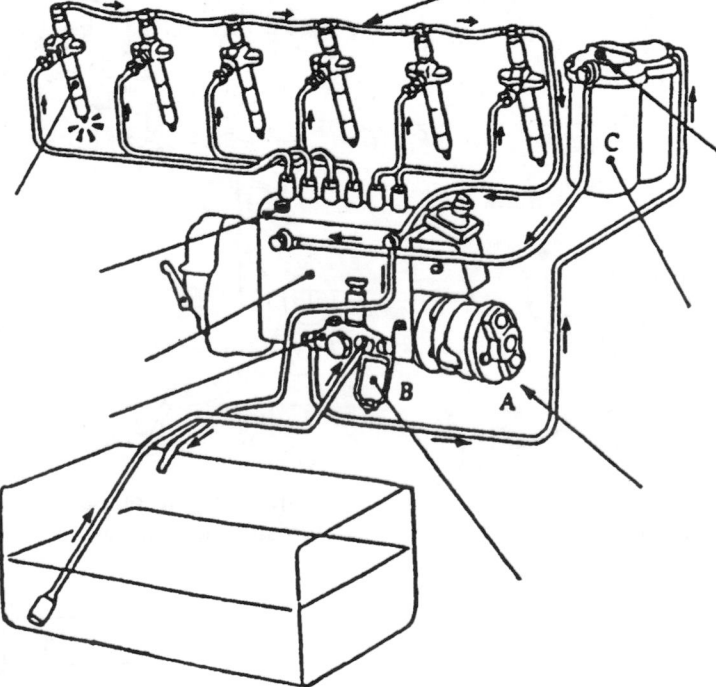

To what is part A connected? ....

What checks should be made to a diesel fuel system during routine maintenance?

....

....

....

....

Describe how the water trap filter B should be serviced:

....

Describe how filter C should be serviced:

....

When filters B and C have been serviced the system must be bled of air. Describe how this bleeding process is carried out:

....

....

....

....

....

....

....

....

....

....

What effects would occur if air was in the fuel system?

....

....

....

When testing injectors, what safety precautions must be observed?

....

....

....

Give TWO reasons why such precautions are necessary:

1. ....

....

2. ....

....

EXHAUST INSPECTION

An exhaust system should be replaced as soon as a box or pipe leaks. During routine service the condition of the exhaust should be checked. Indicate on the complete exhaust system where leaks are most likely to occur.

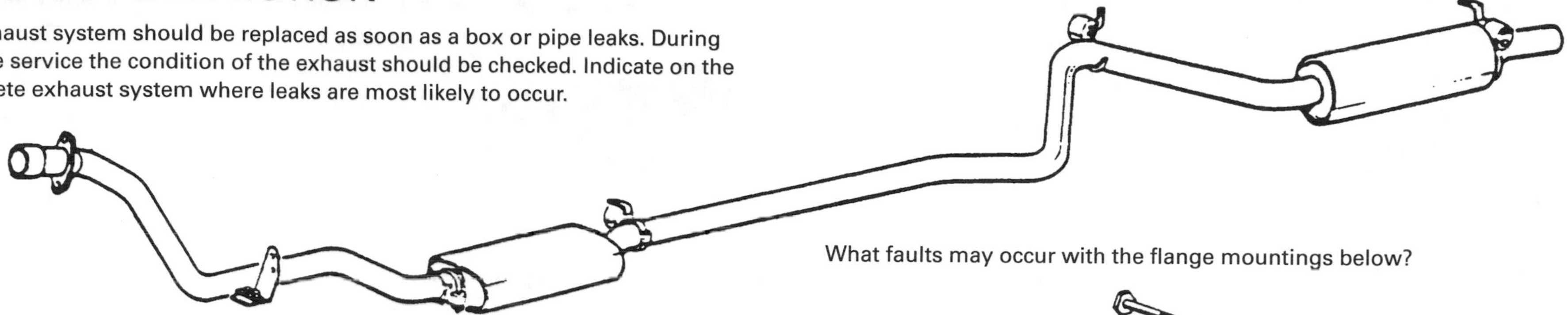

Identify the exhaust supports shown below.

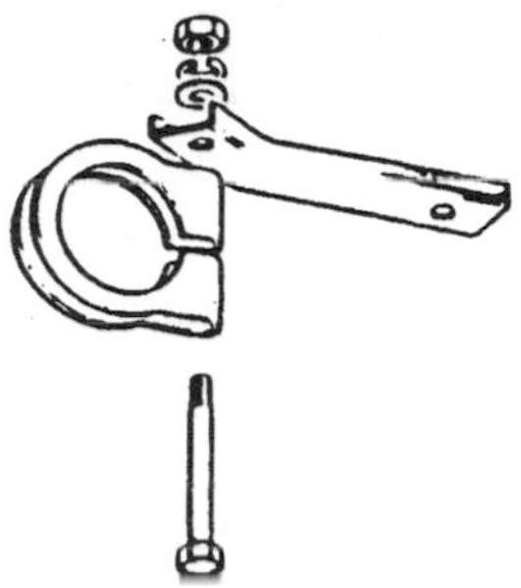

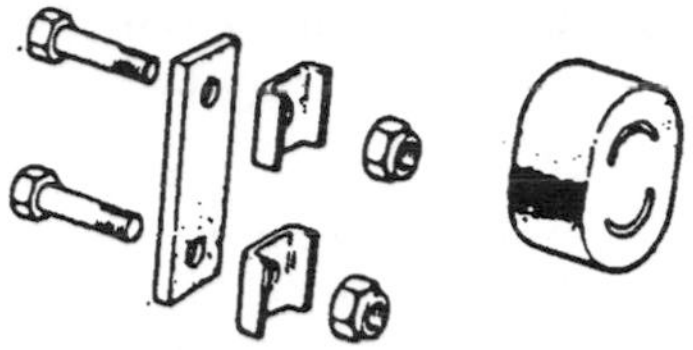

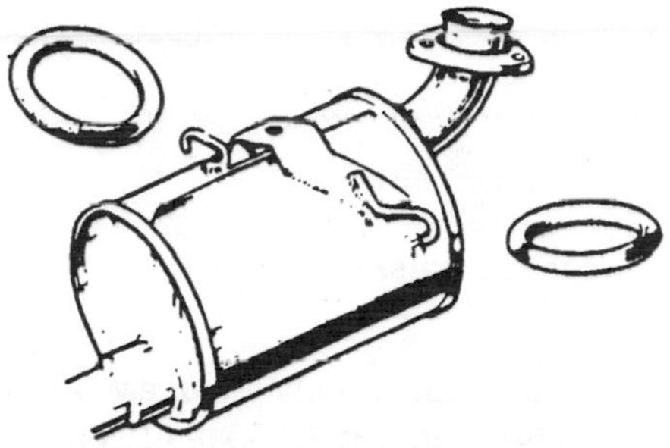

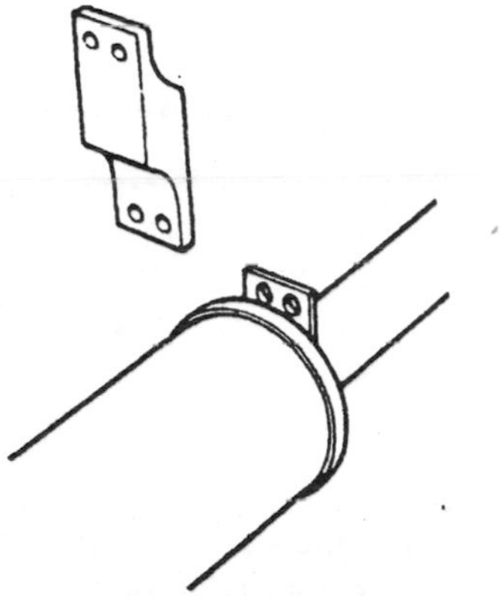

What faults may occur with the flange mountings below?

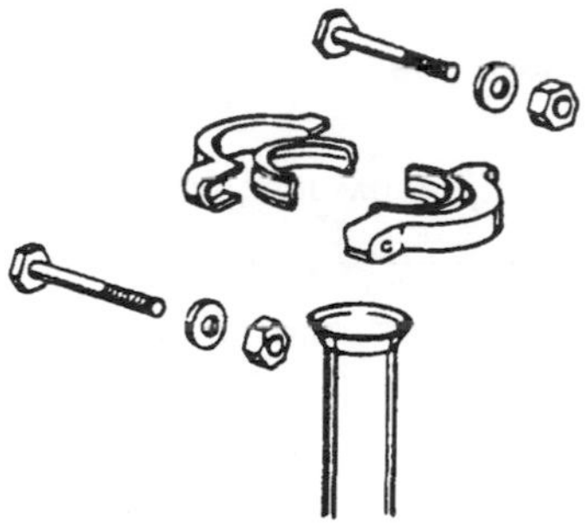

....

....

....

What routine checks should be given to the exhaust system?

....

....

....

....

....

....

....

....

ELECTRICAL SYSTEMS INSPECTION

A basic electrical check will determine the operation/condition of all lamps, horn, wiper and washer systems, the general battery condition and the alternator drive belt tension.

Name the main lighting systems that should be checked:

....

....

....

List FOUR different windscreen wiper operations that should be checked:

....

....

....

....

In what way may wiper blades be faulty?

....

....

The diagram shows, without using specialised equipment, the position of the main beams. What will occur when the lamps are dipped?

....

....

Distance between headlamp centres

Concentrated area of light

Height of lamp centres from ground

Distance for setting is at least 25 feet

Show the approximate dipped position of the concentrated light areas.

The battery needs little maintenance, usually only a visual check, but this should be made regularly.

Complete the table to give common faults that may cause premature battery failure:

Item	Possible fault
Casing	
Top	
Electrolyte	
Connections	
Securing frames	

What faults would make the drive belt unserviceable?

....

....

....

How would the drive belt shown in the diagram be adjusted for correct operation?

....

....

....

....

....

....

....

....

....

....

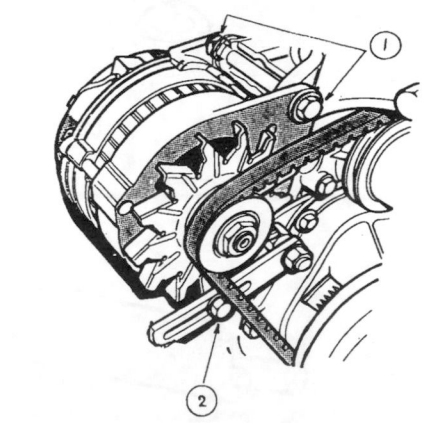

TRANSMISSION INSPECTIONS

Clutch Operation

The drawings show the clutch operating layouts that should be checked during service.

In both cases name the basic parts and state the items that should be periodically checked.

Hydraulically operated

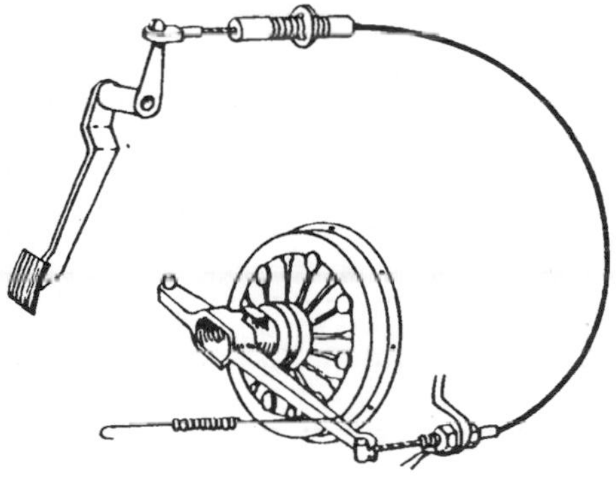

....
....
....
....
....
....
....
....
....

Mechanically operated

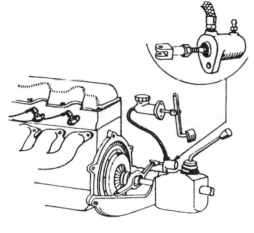

....
....
....
....
....
....
....
....
....

Some clutch gear change layouts on PSVs and commercial vehicles are pneumatically assisted. What checks should be made to these systems?

....

Basic Oil Lubrication

Indicate on the first three drawings from where the oil should be drained and/or checked for the correct level; also indicate from where oil leakage is most likely to occur.

Manual Transmission

Gearbox

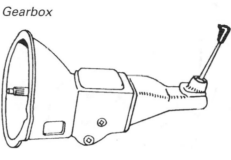

Rear axle

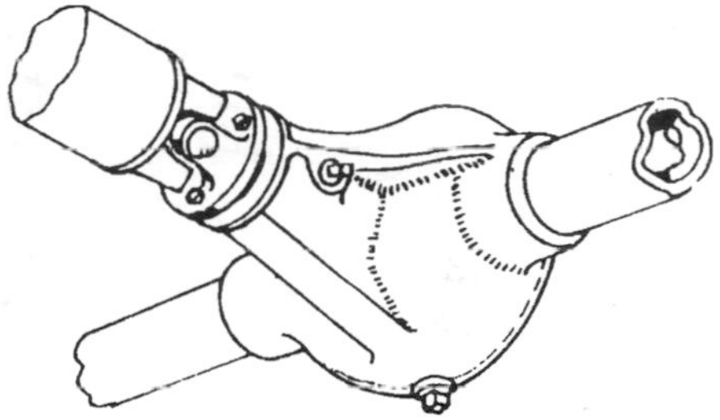

Automatic Transmission

Fluid flywheel

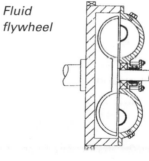

Automatic gearbox dipstick

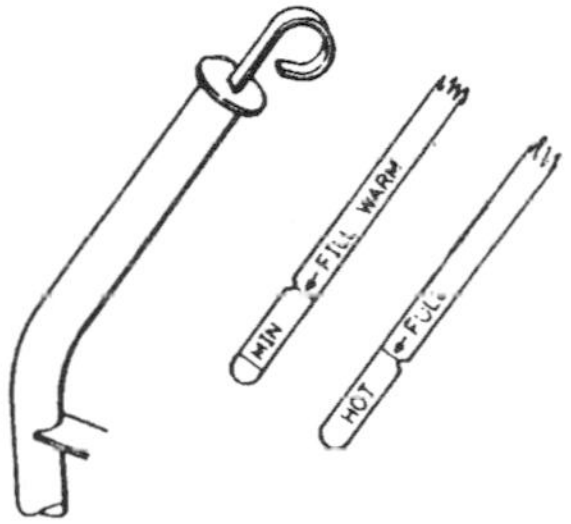

....
....
....

Drive Shafts

What service checks should be made to the drive shaft assembly shown?

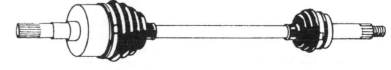

....
....

REAR SUSPENSION INSPECTION

Shown below is a conventional rear wheel drive leaf spring suspension.
What service checks should be made to the numbered items?

....

....

....

....

....

....

....

....

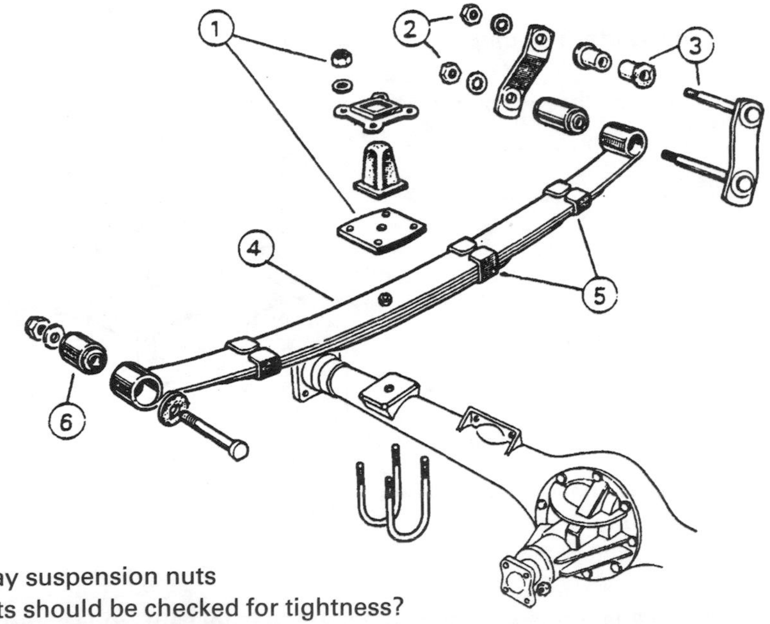

Shown below is a typical rear wheel drive independent suspension system.
Name the numbered items and state what service checks should be made to
them.

....

....

....

....

....

....

....

....

What is the component (5)? ....

What is a typical recommended mileage for its replacement? ....

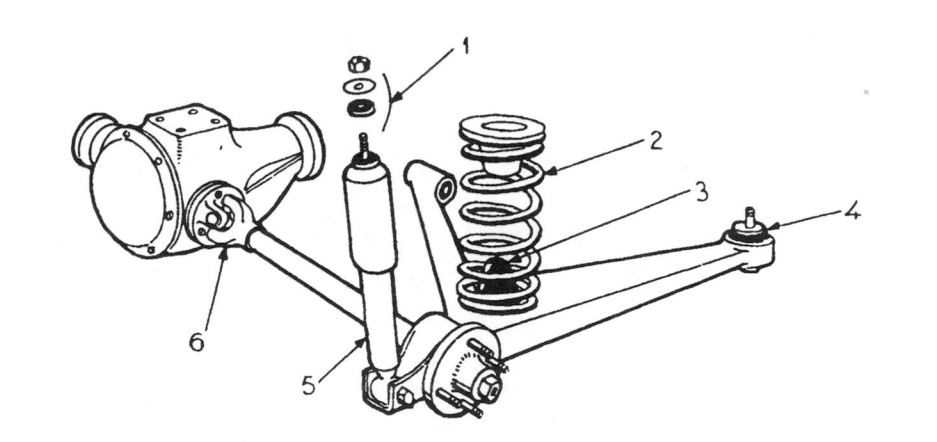

How may suspension nuts
and bolts should be checked for tightness?

....

....

FRONT SUSPENSION INSPECTION

The type of suspension system shown is called:

....

Number the parts and state what service checks should be made to them:

....

....

....

....

....

....

....

....

....

If the ride (trim) height of a vehicle is incorrect (usually it is low), what checks should be made?

....

....

....

....

....

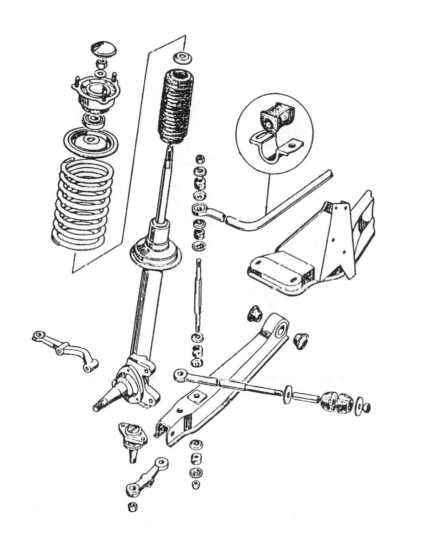

Lubrication

On all the three suspension drawings shown, the moving parts require no lubrication or are self-lubricating.

What parts on older models require lubrication by grease?

....

Some heavy commercial vehicles use an automatic chassis lubrication system by oil or grease.
What periodic maintenance should be given to such a system?

....

....

....

STEERING SYSTEM INSPECTION

The type of steering system shown is the most popular arrangement used on modern cars.

It is called a ....

Indicate and name the items that should be checked for wear, if the steering wheel was rotated to and fro and excessive play was evident.

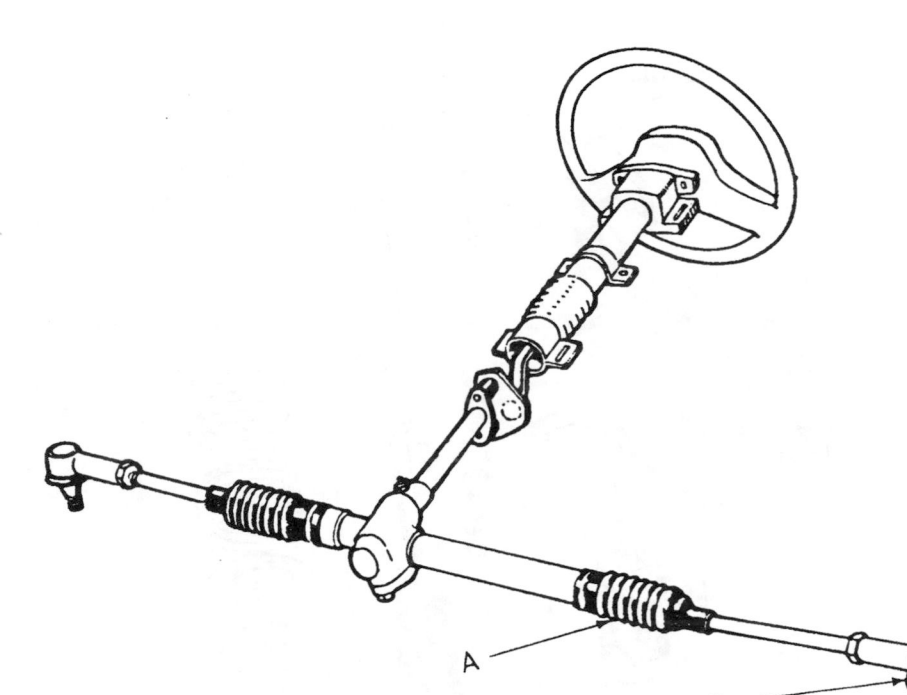

What are items A and B and what checks should be made to them?

....

....

What lubrication is necessary to the steering system?

....

....

Many modern vehicles use a power steering system. Name the main parts of the system layout shown below.

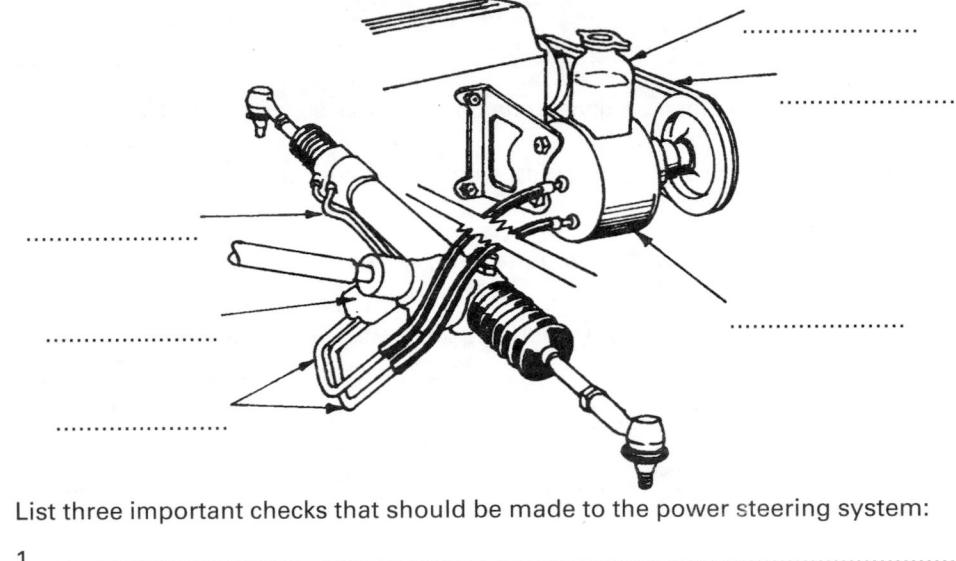

List three important checks that should be made to the power steering system:

1. ....

2. ....

3. ....

Both sides of a filler cap dipstick are shown. What is noticeable about the readings?

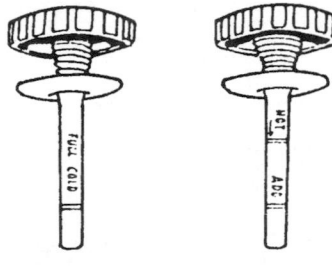

After repair, when the system is topped up with oil, a bleeding procedure must be carried out to expel all air.
Describe how this should be done.

....

....

....

....

....

....

BRAKE SYSTEM MAINTENANCE

Checking a braking system for faults may be divided into TWO sections:

1. Wheels on the ground.   2. Wheels removed.

What initial checks should be carried out with the wheels on the ground?

....
....
....
....
....

What basic checks should be carried out with the wheels removed?

....
....
....
....

Indicate and name the basic parts that require service checks:

Car-dual line (hydraulic) braking system

Checks may also be made to see if the brake indicator warning circuits operate.

To what brake parts are warning indicators generally fitted and what do they monitor?

....
....
....
....

Heavy commercial vehicle brakes require extra attention. State THREE important items to check:

....
....
....
....

Indicate where checks should be made:

Heavy vehicle (air) braking system

Front Disc Brake Assembly

Name the parts that require service checks.

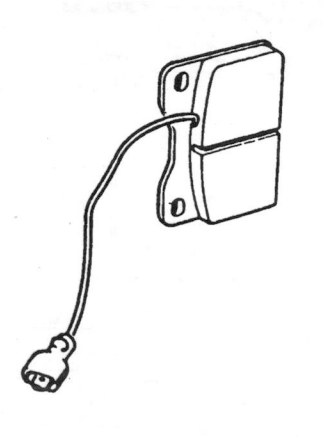
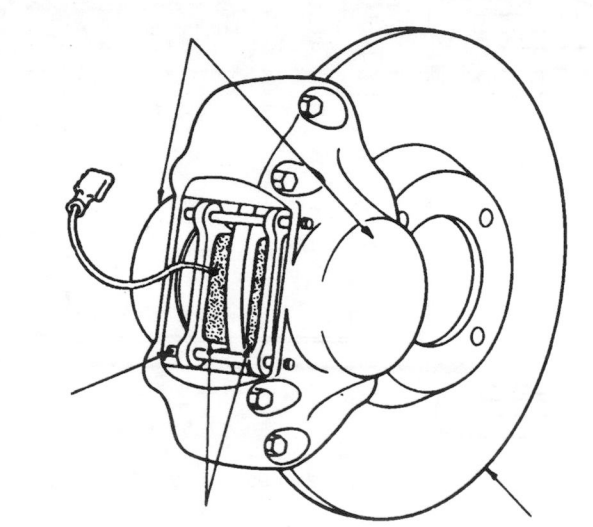

Pedal – Master Cylinder Layout

Name the arrowed parts.
Why is a split reservoir used on the master cylinder?

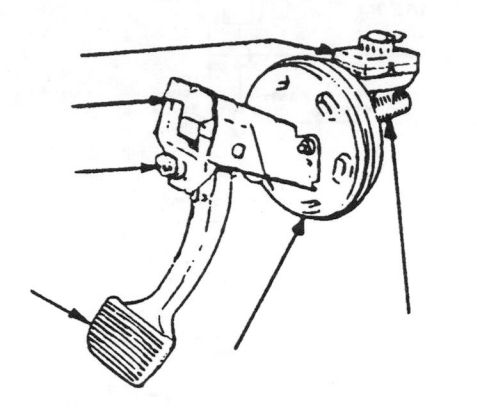

....

....

....

....

....

....

....

State faults that may be expected when checking a disc brake assembly:

....

....

....

....

....

What is the function of the cable hanging from the brake pad shown above?

....

....

....

....

Rear Drum Brake Assembly

Name the parts that require service checks.

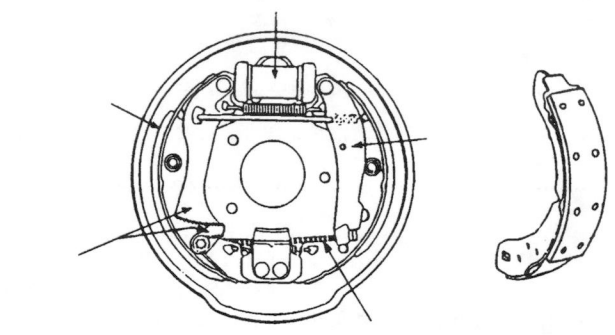

State faults that may be expected when checking a drum brake assembly:

....

....

....

....

INSPECT TYRES

Certain tyre defects make it illegal to use a vehicle on a public road. Visual checks for tyre defects must be made at each service.

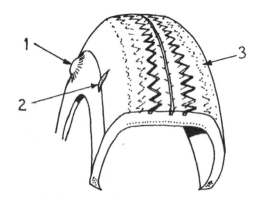

State the three tyre defects shown opposite:

....

....

....

....

....

....

....

....

Types of tyre

Two of the most important characteristics of a tyre are:

(a) type of construction, radial or cross-ply (see Chapter 1), and

(b) its speed rating.

How is it possible to determine these two factors by external examination?

....

....

....

Fitting all cross-ply tyres or all radial-ply tyres to a vehicle is legal; some tyre combinations, however, are dangerous and therefore illegal.

Below are shown three possible tyre fitment combinations. State which is legal and which is illegal.

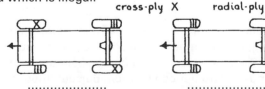

cross-ply X   radial-ply III

....

Ensuring that tyre pressures are correct is probably the single most important tyre check that can be made. Running with tyres incorrectly inflated is both illegal and dangerous.

Compare the recommended pressure for vehicles of the following types:

Type of vehicle	Make	Model	Pressure	
			Front	Rear
Front engine rear drive				
Front engine front drive				
Rear engine rear drive				
Heavy commercial vehicle				

INSPECT ROAD WHEELS

The most important check on a wheel is to ensure that the wheel nuts are tight. Sketch a wheel nut in its correct tightened position and give reasons for its shape.

Hub

....

....

....

State three checks that should be made to road wheels during service:

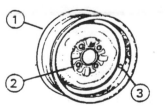

....

....

....

....

....

What other wheel checks should be made?

1. ....

2. ....

INSPECTION AND CARE OF TOOLS

Before commencing work, the condition of hand tools should be checked and any that are damaged should not be used. Similarly, any special service tool drawn from the store should be inspected to ensure that it is safe to use and capable of doing the work required. State examples of when hand tools should be discarded:

....

....

....

....

....

....

Certain test equipment should be checked against known standard readings. Name service items that require testing in such a manner.

....

....

....

....

....

How should tools and equipment be kept in good working condition to ensure that they:

1. Are secured against loss?

....

....

....

2. Are undamaged?

....

....

....

3. Do not deteriorate?

....

....

....

SERVICING SEQUENCE

Every basic service, particularly the larger ones, should follow a predetermined sequence of operations to allow the service to be carried out in the most efficient manner possible.

Produce a list of such a service sequence from collecting the work order to completing the service sheet and report:

1. ....

....

2. ....

....

3. ....

....

4. ....

....

5. ....

....

6. ....

....

7. ....

....

8. ....

....

9. ....

....

10. ....

....

TOOLS USED WHEN SERVICING VEHICLES

Anyone employed in the active repair of motor vehicles should eventually build up a comprehensive set of tools suitable to his specialised type of work.
Name the tools shown and state the sizes or types it is desirable to have:

....

....

....

....

....

....

....

....

....

....

....

....

....

....

....

....

....

....

....

....

....

....

....

....

....

....

....

....

....

List any other basic hand tools you consider desirable:

....

....

Other specialised tools available will most probably be supplied by the garage. What are the most common tools in this area?

....

....

....

What is the basic function of a torque wrench?

....

....

....

What is the purpose of a lock 'C' spanner?

Why is a ring spanner made in the shape shown?

PURPOSE OF SPECIAL TOOLS AND DEVICES USED ON ASSEMBLY

I. *Fitting tools:* Those such as already described are used to secure together various parts or assemblies.

II. *Mechanical joining devices:* These provide the means of joining one component to another and are traditionally nuts, bolts, screws, keys and pins; plus adhesives on modern vehicles. All are adequately covered in other sections but as a reminder, can you identify the items shown?

....

....

III. *Locking devices:* These prevent nuts and bolts from working loose as the vehicle is continually vibrated.
How does the nylon insert in the nut shown provide a self-locking action?.

nylon insert

....
....
....
....
....

Name FIVE other locking devices:

....
....
....
....
....

IV. *Seals:* These ensure that joints are leakproof.
What may the seals be made from on a stationary joint?

....

What type of seal is most commonly used when the component is rotating?

....

V. *Presses:* The types of fit requiring a press for assembly would be ....

A press allows an even, steadily increasing load to be applied – thus eliminating damage from shock loading that, for example, hammering could cause.
State FOUR vehicle components that require assembly by means of a press:

....
....
....
....

VI. *Measuring equipment used during assembly.* In order to ensure an accuracy of fit, before, during or after assembly it may be necessary to take measurements or align the components.

What measuring or testing equipment may be necessary for checking the circumstances listed below?

1. The size or position of a component before and during assembly

....

2. A horizontal or vertical datum

....

3. Angles

....

4. Clearances

....

5. Concentricity

....

6. Position

....

Presses may be operated mechanically or hydraulically. The hydraulic type is usually capable of exerting a higher pressure.

Sketch a press available to you for doing assembly work.

PURPOSE OF SPECIAL TOOLS AND DEVICES USED FOR DISMANTLING

I. *Fitting tools*, such as already described, are used to dismantle components.

II. *Cleaning agents:* What is their purpose before, during and after dismantling?

....

....

....

III. *Dyes and markers:* Why are these used?

....

....

....

What type of markers are available?

....

....

....

IV. *Stud removal:* The sketches show a stud remover and a stud extractor. Identify them and state how a stud remover should be used:

....

....

....

....

....

....

....

....

V. *Penetrating oil:* What is its function?

....

....

VI. *Special-purpose tools:* These are made to do a specific job on a vehicle and are numerous and varied. They range from a spark plug spanner to inner door handle removal tools.
Give four examples of such special tools:

....

....

....

....

VII. *Gear or bearing pullers:* Many items require a puller to remove safety a gear or bearing from a shaft. These pullers may be purpose-made for a specific job or be a universal puller with attachments making it suitable for many jobs.
Complete the sketch of the bearing puller shown below.

VIII. *Presses:* If a bearing or flange is too tight to be removed by a puller, the whole unit may require removing from the vehicle and placing in a press. What is the advantage of this?

....

....

Note: Every attention must be given to safety when the load is applied.

IX. *Heating equipment:* When all else fails to remove a component, heat can be applied by an oxy-acetylene flame. How will this release the part?

....

....

DANGERS IN ASSEMBLY AND DISMANTLING

When assembling or dismantling, safety precautions must be observed at all times. A lack of concentration could cause an accident to yourself or damage to the component.

State some important precautions to be observed under the following headings:

Forces – weight of system and forces applied	Engine removal ....
	....
	Jacks and stands ....
	....
	Fuel tanks ....
Pressure and flammability	Air brakes ....
	Petrol tanks and pipe ....
	....
	Garage pipework ....
Electricity	Battery ....
	....
	Mains electricity ....
Temperature	Cooling systems ....
	....
	Oil changing ....
Chemicals	Battery, cooling system, brake system ....
	....
	....

| Component marking for re-assembly | .... |
| | .... |

GENERAL RULES FOR ASSEMBLY AND DISMANTLING

The workshop manual provides a dismantling and assembly procedure which should be used at all times until the correct method is known.

State general rules that should be applied to ensure that successful dismantling and assembly occur:

1. ....
....
....

2. ....
....

3. ....
....

4. ....
....
....

5. ....

6. ....
....

7. ....

WORKSHOP JOB SHEET

With the aid of information from your firm regarding an actual job that you have carried out, complete the workshop job sheet on the next page. Ensure that you understand all the items that are mentioned.

Workshop Job Sheet

Job No. ....

Mechanic's/Trainee's name

....

Date ....

Vehicle Owner's Name/Customer

....

Address ....

....

....

.... Phone No. ....

Vehicle make ....

Model ....

Reg. No. ....

Mileage ....

Vin No. ....

Engine No. ....

MOTOR VEHICLE NVQ

RECORDED/EVIDENCE

OF WORK COMPLETED

Work Instructions	Repair Times	TYPE OF SYSTEMS CHECKED (✓ Box(es) Applicable)	Time taken	Charge rate

Work Instructions

....

....

....

Any items found that require Extra Work

....

....

....

Repair Times

....

....

....

....

....

....

....

TYPE OF SYSTEMS CHECKED
(✓ Box(es) Applicable)

Engine	☐	S I Fuel	☐
C I Fuel	☐	Cooling	☐
Ignition	☐	Transmission	☐
Brakes	☐	Suspension	☐
Steering	☐	Wheels/Tyres	☐
Starter	☐	Batt. & Charge	☐
Elect. Aux.	☐	Body Fittings	☐
Security	☐	Augmentation	☐

Time taken
....
....
....
....
....
....
....

Charge rate
....
....
....
....
....
....
....

Work Carried Out

....

....

....

....

....

....

....

Materials used: Description

....

....

....

....

....

....

....

Cost

....

....

....

....

....

....

....

Time started:

Time finished:

Portfolio No.

Attach to this sheet any copies of technical information used. E.g. Manufacturers' Check Sheets, Engine Diagnostic/Brake Tester/ Exhaust Gas Analyser - Readouts.

Customer's signature

Assessor's/ Supervisor's signature

Method of Payment:

Warranty	☐	Cash	☐	F O C	☐
Account	☐	A/C No.			
Credit Card	☐	CC No.			
Voucher	☐	V No.			

Totals
Not VAT Items
FINAL
TOTAL

....

....

FAULT DIAGNOSIS

BASIC PRINCIPLES

All motor vehicle faults can be found by systematic testing. Some faults may be so obvious that they can be found with little investigation, e.g. an exhaust making a very loud noise. Other faults, like an engine misfiring intermittently may prove much more difficult to solve. Diagnosis by systematic testing means that a similar approach can be made to solve any fault.

List a typical 6 Point procedure for solving vehicle faults by diagnosis.

1. ....
....
2. ....
....
3. ....
....
4. ....
....
5. ....
....
6. ....
....

Use the above procedure to describe how to diagnose the faults suggested on this and the next page. To solve many problems you will need to use specific manufacturers technical data.

ENGINE AREA

Customer states 'The oil pressure light comes on when the engine is running slowly but the engine still performs satisfactorily.'

From this information the probable fault is:

....

Diagnostic test procedure

1. ....
....
....
....
....
....
....
....
....
....
....

Suitable test equipment.

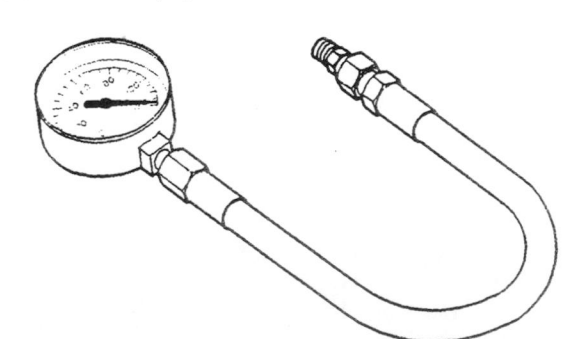

Engine oil pressure testing gauge

ENGINE AREA

Customer states 'The engine temperature is high and the coolant level is low.'

From this information the probable fault is:

....

Diagnostic test procedure

1. ....
....
....
....
....
....
....
....

Suitable test equipment.

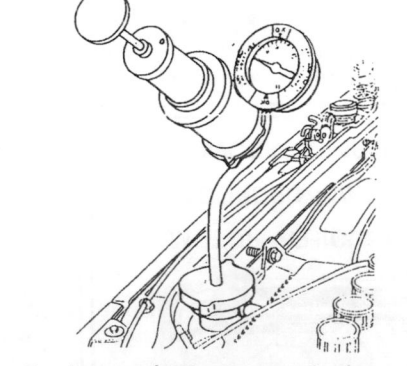

Engine coolant pressure tester

TRANSMISSION AREA

Customer is concerned because when the car goes up a hill the engine revs faster but the vehicle does not accelerate.

From this information the probable fault is:

....

Diagnostic test procedure

1. ....
....
....
....
....
....
....
....
....
....
....
....

Suitable test equipment.

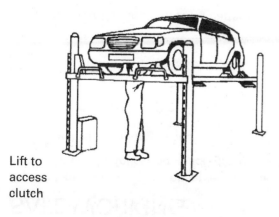

Lift to
access
clutch

CHASSIS OR FRAME AREA

Customer complains that the tyre treads on the front wheels are wearing unevenly.

From this information the probable fault is:

....

Diagnostic test procedure

1. ....
....
....
....
....
....
....
....
....
....
....

Suitable test equipment.

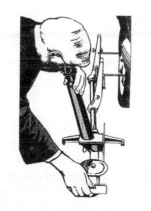

Optical
Alignment
Gauge

ELECTRICAL AREA

Customer states that 'A red light on the dash is on all the time and the starter is cranking slowly.'

From this information the probable fault is:

....

Diagnostic test procedure

1. ....
....
....
....
....
....
....
....
....
....
....

Suitable test equipment.

Multimeter

SAFE WORKING

State the precautions to be observed to avoid injury or vehicle damage when carrying out scheduled servicing.
Itemise under the following headings:

Jacking vehicle, working in pits

....

....

....

....

Working under bonnet, removing battery

....

....

....

....

Running engine

....

....

....

Brake servicing

....

....

....

....

Basic servicing

....

....

....

Car body and interior protection

....

....

....

....

State the safety faults depicted by the drawings:

Tools

Electrical

Wearing jewellery

Making vehicle wheel free

Using pits

Parts storage

Chapter 7

Measurement and Dimensional Control

SIMPLE MEASUREMENT AND DIMENSIONAL PROPERTIES

Many workshop processes, e.g. manufacturing, construction or repair operations, involve some form of measuring. Measuring is done by COMPARING the dimension or object to be measured with a similar object or some kind of measuring tool or instrument.

Give TWO simple examples of how measurement information can be obtained by means of a simple COMPARISON process:

1. *Where, say, a series of bolts needs to be replaced, the new ones can be* ....

   easily checked against the existing bolts. ......

2. ....
   ....
   ....
   ....

Measuring equipment may be NON-INDICATING or INDICATING; describe briefly the difference between the two types of equipment.

....
....
....
....
....

Name the measuring instruments shown below and state whether they are indicating or non-indicating.

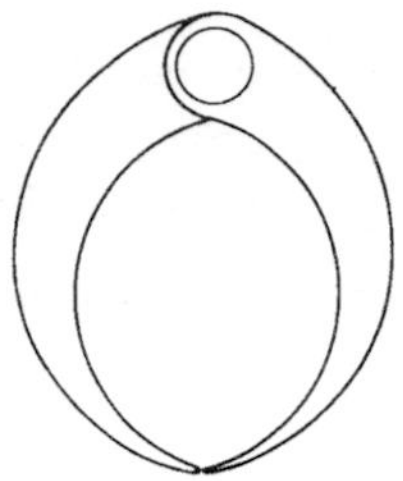

....
....

....
....

....
....

DIMENSIONAL PROPERTIES

Some important dimensional properties which can be measured are

(a) *LENGTH (width, height, depth, diameter, radius)* ....

(b) ....   (c) ....

(d) ....   (e) ....

(f) ....   (g) ....

How might you check the surface of a cylinder head for FLATNESS in garage conditions?

....
....
....
....

For the two road wheels shown at (a), measurements taken at x and y would give an indication of:

....

Which measuring instrument could be used to determine the 'relative position' of the road wheel shown at (b) to the vertical plane?

....

(a)   **(b)**

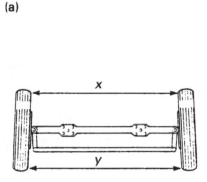

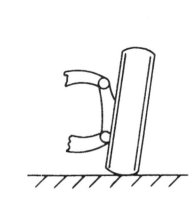

STANDARDS

Standards of measurement and dimensional control have been agreed on a national and international basis. These standards relate to:

(a) *Quantities and units used in measurement.* ....

....

....

....

The International Organisation for Standardisation (ISO) sets the standards relating to measurement and dimensional control on an international basis.

Who is the recognised authority for the preparation and publication of standards in the UK?

....

The SI system (Système International) is an international system of units set up to standardise quantities and units used in measurement on an international basis. Some quantities used in the SI system are listed below; state the correct SI unit for each quantity (and its correct abbreviation).

Quantity	Unit	Abbreviation
Length		
Time		
Mass		
Velocity		
Force		
Pressure		

Quality specifications for tools and measuring equipment

If standards applied to measurement are to be achieved, the tools and equipment used for measurement must have a higher degree of accuracy than those required for dimensional properties being measured.

To maintain the standard of accuracy required, tools and equipment must be HANDLED, MAINTAINED and STORED with great care.

SYMBOLS OF BASIC QUANTITIES IN THE SI SYSTEM

All the names of quantities commonly used in motor-vehicle technical data are given abbreviated symbols. This is so that the full name of the quantity need not be written out each time it is used. The symbols for the units have already been used on the previous page. When shown in data books the symbols for units are printed in ordinary print and the symbols for quantities are printed in italics, e.g. (metre = m) (mass = *m*).

Complete the symbols of the quantities and units shown.

Quantity	Quantity symbol	Unit of measurement	Unit symbol
Length	....	....	....
Mass	....	....	....
Time	....	....	....
Area	....	....	....
Volume	....	....	....
Speed (velocity)	....	....	....
Acceleration	....	....	....
Force	....	....	....

MULTIPLES AND SUB-MULTIPLES OF SI

In the SI system, prefixes are used in front of the basic unit to eliminate the need for writing many noughts on large or small numbers.

Complete the table below which shows those prefixes.

Prefix	Symbol	Factor	Value written in full
giga	....	....	....
mega	....	....	....
kilo	....	....	....
hecto	....	....	....
deca	....	....	....
basic unit			
deci	....	....	....
centi	....	....	....
milli	....	....	....
micro	....	....	....
nano	....	....	....

It is preferred that numbers are expressed in units between 0 and 999, e.g.

$$50 \text{ m} = 50 \text{ m}$$
$$50000 \text{ m} = 50 \times 10^3 \text{ m} = 50 \text{ km}$$
$$0.05 \text{ m} = 50 \times 10^{-3} \text{ m} = 50 \text{ mm}$$

In practice the basic unit and the milli, kilo and mega units are very commonly used.

The values centi, deci, deca and hecto are only used when dealing with areas and volumes where it may be more convenient to use them because square and cubic millimetres become awkward to use, e.g.

$$1 \text{ m}^3 = 1\,000\,000\,000 \text{ mm}^3 = 10^9 \text{ mm}$$

Complete the following statements.

The number of metres in a kilometre is ....

The number of millimetres in a kilometre is ....

The number of newtons in a meganewton is ....

The number of milligrams in a gram is ....

The number of watts in a gigawatt is ....

Change each of the following values to its preferred value.

1. 7800 m
   ....
2. 0.006 m
   ....
3. 0.000 06 m
   ....
4. 22 000 g
   ....
5. 5 200 000 g
   ....

6. 200 000 N
   ....
7. 9 673 000 N
   ....
8. 3600 MW
   ....
9. 430 cm
   ....
10. 0.0087 A
    ....

ENVIRONMENTAL STANDARDS

Because most materials expand when subjected to an increase in temperature, an agreed environmental temperature of .... is used when calibrating and setting up measuring tools and equipment.

What is the purpose of the STANDARDS ROOM?

....

....

....

....

....

State the advantages of having agreed standards.

1. *Interchangeability of component parts.* ....

....

....

....

....

MEASUREMENT OF LENGTH

The equipment used for measuring length depends on:

(a) *The actual purpose for which it is to be used.*

(b) ....

Frequently used instruments for measuring length are shown below.

(i) Name the instruments and (ii) state the degree of accuracy to which they can be used:

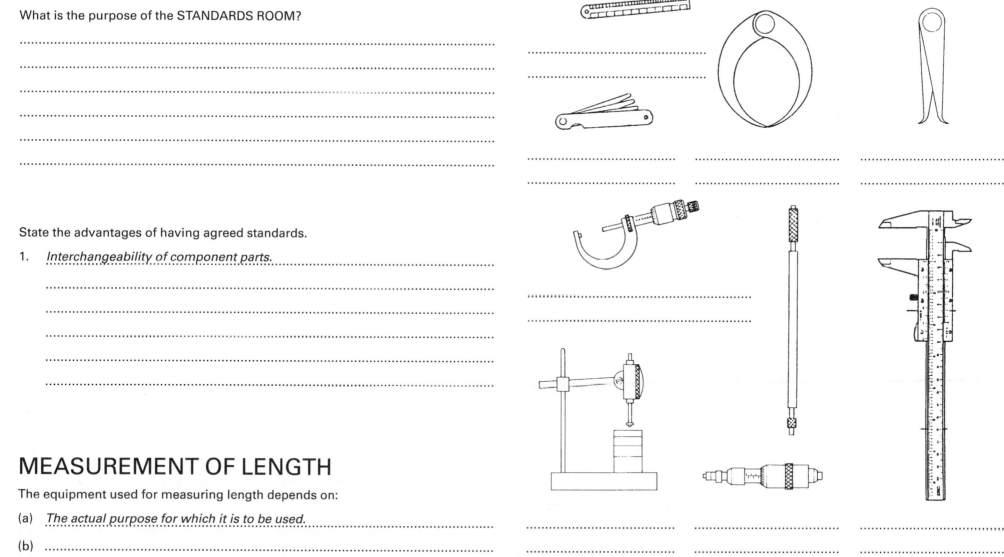

....

....

....

....

....

....

....

....

....

....

....

....

....

....

....

The scientific unit of measurement for length is the metre. A simple measuring instrument is the rule. This is used for most general purposes. If an accuracy to a limit of 0.01 mm is required, then a vernier or micrometer caliper may be used.

Using a rule measure the length of the lines below. State answer in units required.

Draw accurately lines to the requirements below.

_____	.... mm 5 mm
_____	.... mm 17 mm
_____	.... mm 32 mm
_____	.... mm 0.04 m

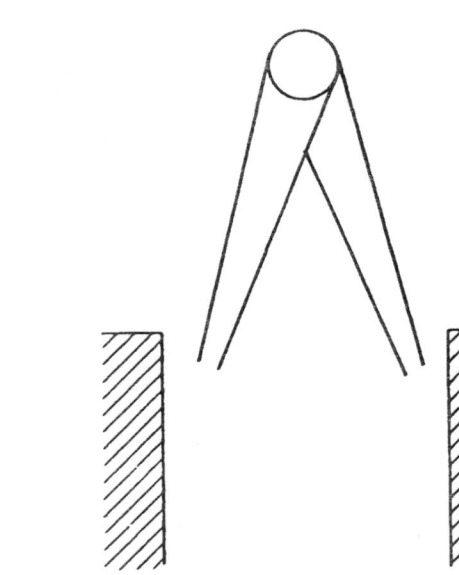

Sketch the feet of the external caliper being used to obtained the diameter of the shaft.

Sketch the feet of the internal caliper being used to obtain the diameter of the bore.

MICROMETER

The micrometer caliper is used to measure the diameter of components to an accuracy of 0.01 mm (1/100 mm).

The principle on which it operates to achieve this accuracy is described below:

The screw connecting the micrometer spindle to the thimble has a lead of 0.5 mm and the barrel divisions are scribed 0.5 mm apart. Therefore, one revolution of the thimble moves the spindle and thimble exactly one barrel division (0.5 mm). The barrel divisions are placed on either side of the datum line. The thimble has 50 equal divisions therefore the movement of 1 thimble division will give an accuracy of 0.5/50 = 0.01 mm.

Micrometers are made in size ranges of 0–25 mm, 25–50 mm, 50–75 mm, etc. A big end bearing journal of 45.50 mm diameter would be measured by using a 25–50 mm micrometer.

Examine a micrometer and complete the sketch of the one positioned measuring the bar below. Name the important parts.

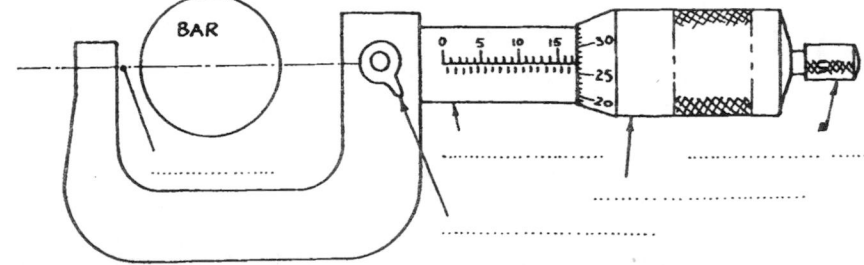

Micrometer reading ....

Reading the micrometer scale

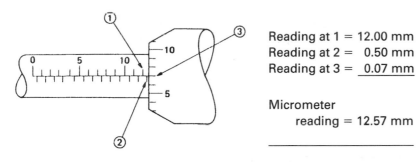

Reading at 1 = 12.00 mm
Reading at 2 = 0.50 mm
Reading at 3 = 0.07 mm

Micrometer
reading = 12.57 mm

State the readings of the scales of the metric micrometers below.

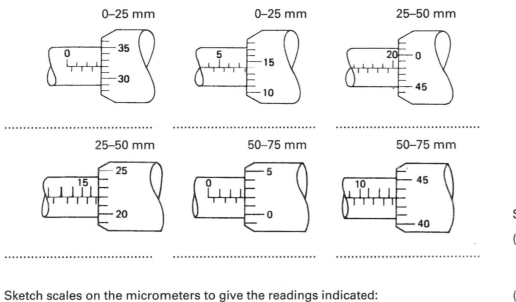

0–25 mm

0–25 mm

25–50 mm

25–50 mm

50–75 mm

50–75 mm

Sketch scales on the micrometers to give the readings indicated:

8.17 mm   14.76 mm   5.22 mm

VERNIER CALIPER GAUGE

The degree of accuracy of a measuring instrument depends upon the fineness of the divisions marked on its scales. The metric rule gives an acceptable degree of accuracy up to 0.5 mm while the vernier caliper gauge can provide an accuracy of 0.02 mm (*Note:* 0.02 mm is a smaller accuracy than 0.001 in.)

Name the arrowed parts of the vernier caliper shown below:

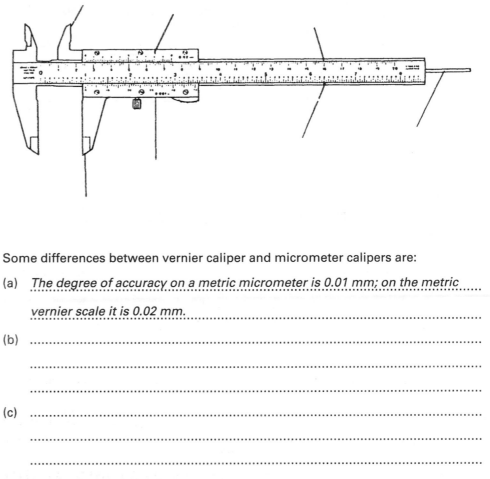

Some differences between vernier caliper and micrometer calipers are:

(a) *The degree of accuracy on a metric micrometer is 0.01 mm; on the metric vernier scale it is 0.02 mm.*

(b) ....

....

(c) ....

....

....

PRINCIPLE OF READING SCALES

The vernier caliper consists of two slightly different scales, one fixed and one moving.

These gauges may be found calibrated in both English and metric scales.

There are two types of metric scales, in both cases the reading on the sliding scale is multiplied by 0.02.

(a) 25-division vernier (sliding) scale, the main scale is graduated in 0.5 mm divisions.

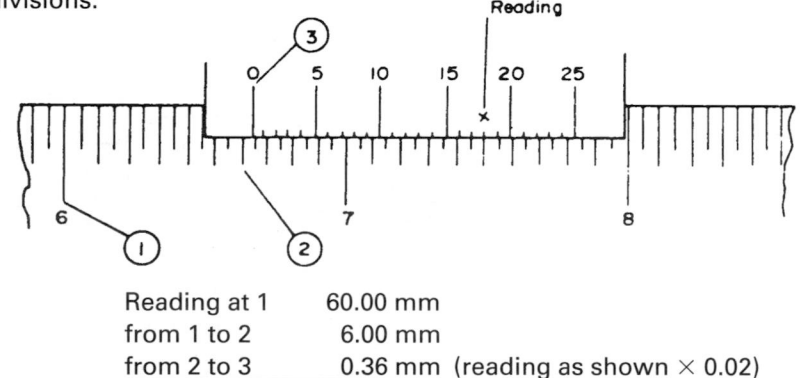

Reading at 1	60.00 mm	
from 1 to 2	6.00 mm	
from 2 to 3	0.36 mm	(reading as shown × 0.02)
reading	66.36 mm	

(b) 50-division vernier (sliding) scale. The main scale is graduated in 1 mm divisions.

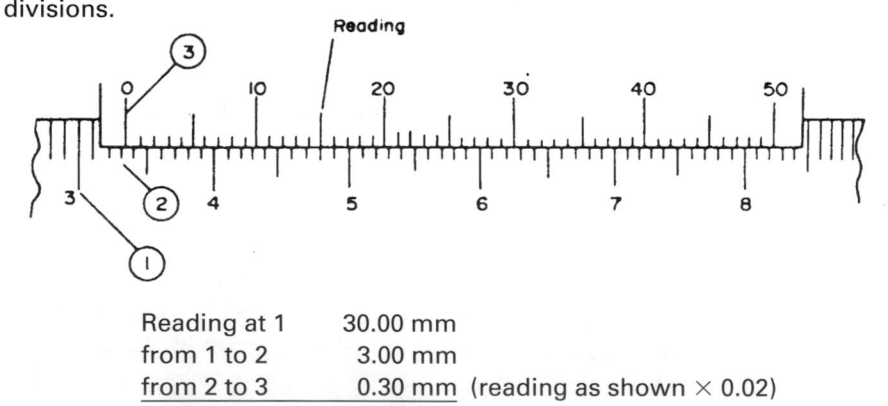

Reading at 1	30.00 mm	
from 1 to 2	3.00 mm	
from 2 to 3	0.30 mm	(reading as shown × 0.02)
reading	33.30 mm	

State the readings on the scales below:

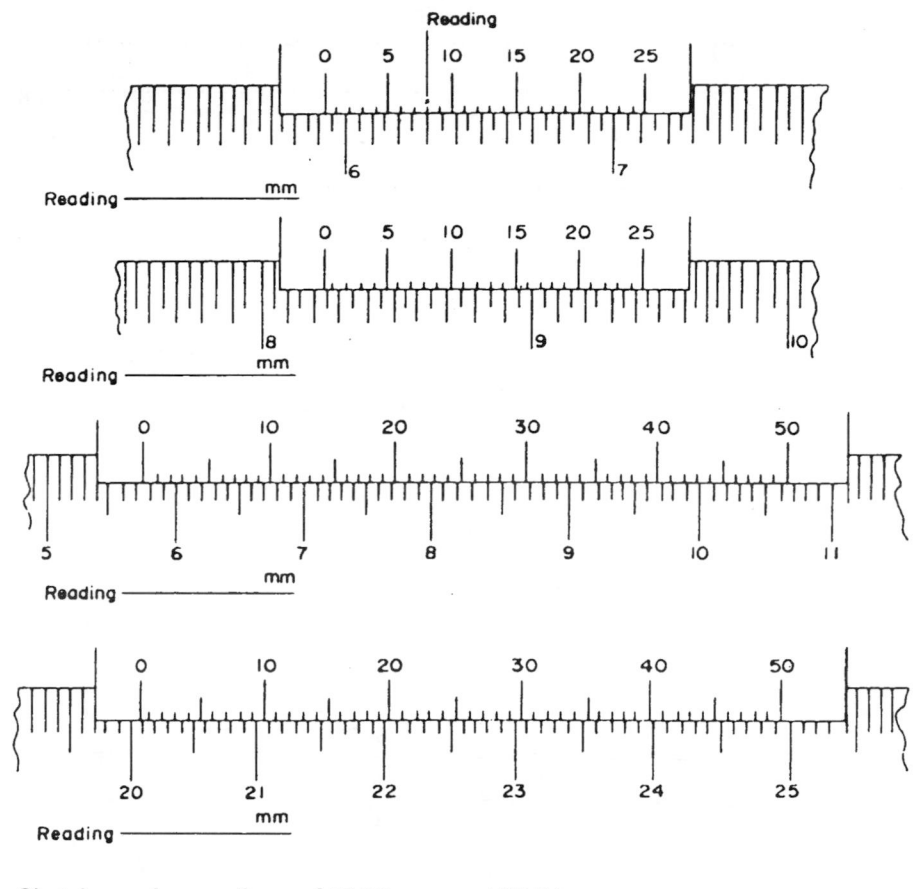

Reading ———————— mm

Reading ———————— mm

Reading ———————— mm

Reading ———————— mm

Sketch vernier readings of 17.86 mm and 53.24 mm:

DIAL INDICATORS

A dial test indicator (DTI) is an instrument which may be used to give comparative measurements from an item of specific or standard size. The main type of indicator used in motor-vehicle repair work is the plunger type. Slight upward pressure on the plunger moves it upwards and the movement is indicated by the dial finger.

A typical clock gauge with a measuring range of 10 mm would have an accuracy of 0.01 mm.

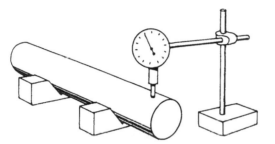

How would the dial indicator be used to check the shaft shown above for roundness or concentricity?

....

....

Give three examples of the uses of a DTI in motor-vehicle repair:

1. ....

2. ....

3. ....

   ....

The cylinder bore gauge

An internal micrometer takes direct readings but requires a sensitive touch to obtain accurate readings. When a cylinder bore gauge is used, it must be calibrated by using either a ring gauge or an external micrometer.

The cylinder bore gauge converts the horizontal movement of the spring-loaded plunger into a vertical movement, which is transferred by a push rod (in the gauge shaft handle) to a dial test indicator clamped to the top of the handle.

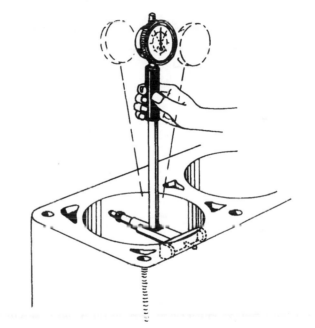

Why is the gauge being rocked to and fro as shown above?

....

....

....

....

ANGLES AND THE UNIT OF ANGLE

Angles are formed when a circle is divided radially into parts. The unit of angle is obtained by dividing the circle into .... equal parts and 1/360 of a circle is a .... which is the UNIT OF ANGLE. The circle shown is divided into four equal parts whose angles are .... Divide the circle further, without using measuring equipment, to show approximate angles of 45°.

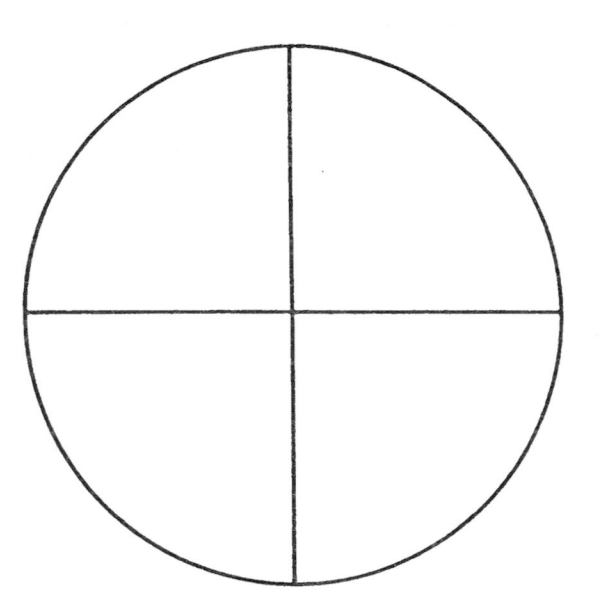

Each degree can be divided into 60 equal parts:

1/60 of a degree = 1 ....

or 1 degree   = 60 .... and

if 1′ is further divided by 60 the ....

is obtained.

1′ = 60 ....

Angular Measurement

Angles and squareness can be checked using:

1. *Try squares* ....

2. ....

3. ....

The try square is the most common tool used for checking squareness. Show by simple sketches how the try square would be used to check the angles (a) and (b) for squareness on the workpieces shown below.

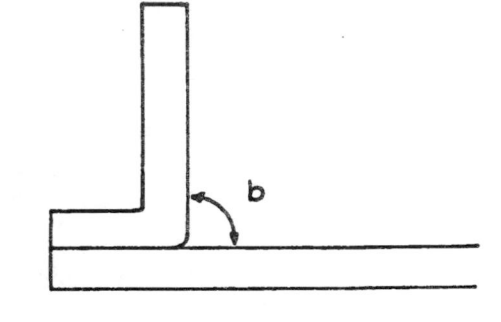

When measuring angles other than 90°, a protractor can be used; the type of protractor depends on the particular application and the degree of accuracy required for the measurement.

Bevel Protractor or Combination Square

A useful instrument for measuring angles is the universal bevel protractor or combination square. A combination square consists of a blade which may be used with any one of three heads; these are illustrated opposite.

State the purpose of a 'centre square' and describe how it would be used. Show by sketching how the workpiece would be positioned when using the centre square opposite.

....
....
....
....
....
....
....

Indicate on the protractor opposite where the angle being checked is read.

Vernier protractor

The vernier protractor is a protractor with a vernier scale which can be adjusted to give a higher degree of accuracy when measuring angles; however, using it requires skill and experience.

How does the accuracy of a vernier protractor compare with that of the bevel protractor shown below?

Normal bevel protractor accuracy = 0.5°

vernier protractor accuracy = 0.5′

Square

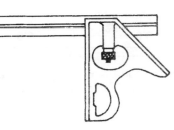

Centre square

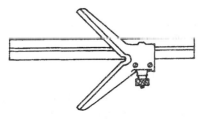

Protractor

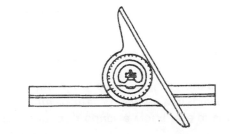

SURFACE TABLES AND SURFACE PLATES

Surface tables and plates both fulfil the same purpose in a workshop. The surface plate is the smaller of the two and more portable. Both are to be found in a mechanical engineering workshop, whereas in a motor-vehicle repair workshop a surface plate is quite adequate.

Surface tables and plates are used for marking out and checking workpieces and components. They are an extremely rigid cast iron structure. Their main feature is a highly accurate planed surface which provides a stable accurate datum base for marking out and checking workpieces and components.

Spirit level

Its function is to ....

....

Sketch a typical spirit level in the space below.

Note: See Chapter 9 for other associated tools.

DIMENSIONAL DEVIATION

It is not possible to measure or produce a workpiece or component *exactly* to size or to duplicate *absolute* accuracy between workpieces. When components are being manufactured, therefore, a DEVIATION from a specified dimension is allowed.
A standard system of SIZE LIMITS controls the dimensional accuracy.

Definitions

Nominal size
This is the design size, and if it were possible to work to an exact size a ....

component would be made to the nominal size. ....

Limits ....

....

....

....

Tolerance

....

....

Deviation

....

....

State the nominal size, high and low limits, tolerance and dimensional deviations for the following dimensions.

	$50 \begin{array}{l} +0.5 \\ -0.5 \end{array}$	$25 \begin{array}{l} +0.5 \\ -0.2 \end{array}$	$30 \begin{array}{l} +0.6 \\ -0.0 \end{array}$
Nominal Size			
High Limit			
Low Limit			
Tolerance			
Deviation			

ACCURACY IN MEASURING

The degree of accuracy is directly related to the tolerance. If high accuracy is not required for a particular component, the manufacturing tolerance can be greater. The choice of measuring equipment for a particular job depends on the tolerances being worked to. For the tolerances given at (a), (b) and (c), name the measuring equipment to be used.

	Tolerance	Equipment
(a)	0.1 mm	....
(b)	0.25 mm	....
(c)	0.04 mm	....

Factors Affecting Accuracy

Give some examples of things which can affect accuracy when taking measurements:

(a) *Variations in temperature affecting measuring tools and workpieces.*
....
....
....
....
....
....
....

Many errors in measurement are caused by incorrect use of the measuring equipment.

Some common causes of error are shown below; make simple sketches to illustrate correct measuring techniques.

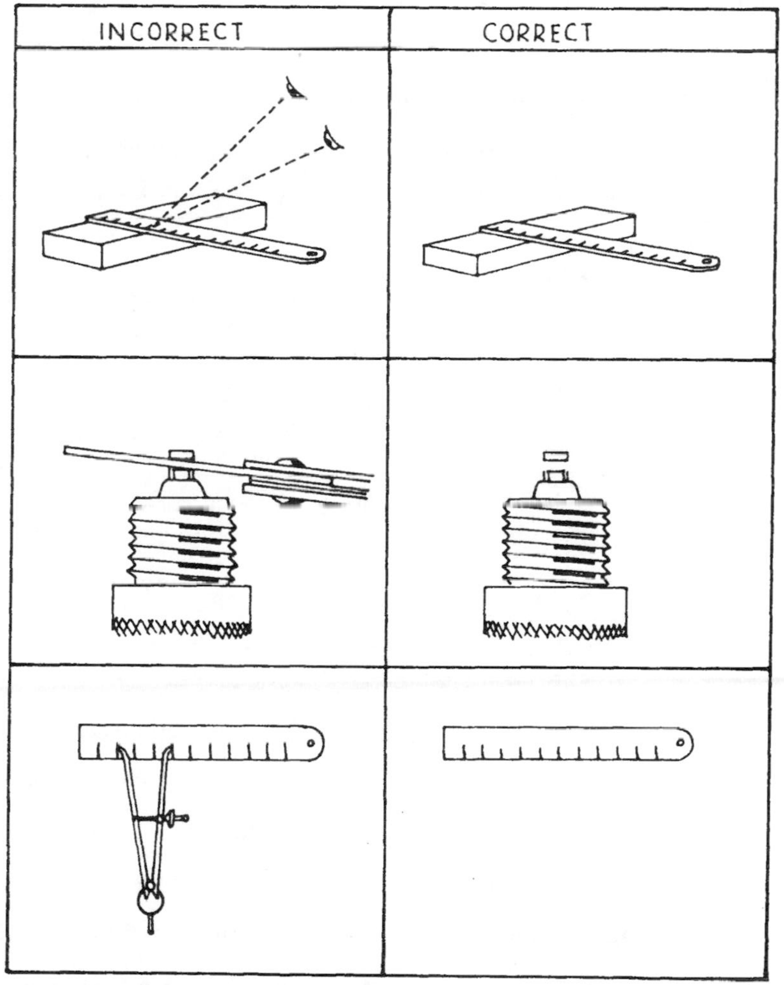

203

TERMINOLOGY

Define the following terms:

(a) Measuring range

....

....

(b) Reading value

....

....

(c) Indicated size

....

....

(d) Reading

....

....

....

(e) Mean size

....

....

The accuracy of the measuring equipment depends on its *condition* and *quality*. A measuring tool may indicate a size which can deviate from the dimension being measured.

Define *tool accuracy*.

....

....

....

CARE AND MAINTENANCE OF MEASURING EQUIPMENT

Keeping measuring tools and equipment in good condition depends on:

(1) the way in which they are handled and used in the workshop;

(2) how the equipment is stored when not in use;

(3) the maintenance of the equipment.

List some important points with regard to the upkeep of:

Rules

....

....

Squares/Feeler gauges

....

Micrometers/Verniers/Dial indicators

....

....

....

....

Very often measuring tools and equipment are damaged as a result of misuse, e.g. using rules, feeler gauges, squares for purposes other than measuring.

Measuring equipment should always be:

....

....

....

Chapter 8

Interpreting Drawings, Specifications and Data

DRAWINGS, SPECIFICATIONS AND DATA

Engineering drawing is a means of communication. It is a simpler, more accurate and less ambiguous form of communication than the spoken or written word. By 'reading' and understanding a drawing, the engineer can determine:

(a) the shape and dimensions of a component
(b) the constructional features, layout and location of components.

How could (a) and (b) above be of use to a motor vehicle technician?

(a) ....

....

....

(b) ....

....

....

....

....

Information Standards

To understand drawings it is necessary to learn the 'language' and the simple rules of engineering drawing. Although motor engineers mostly need to read workshop manual drawings, it is useful to know something about engineering drawing.

To enable all engineers to understand and use the same rules when drawing, British Standard 7308 is used.

Look through BS 7308 (Student Edition) and list the main items of content:

....

....

....

....

....

....

....

DIN, SI and SAE are abbreviations for standards/conventions relevant to the transport industry.

DIN

Deutsche Industrie Norm; the German equivalent of BSI, it is concerned with the harmonisation of industrial standards.

SI

Système International d'Unités; an international system of units which is a modification of the metric system consisting of six base units.

SAE

Society of Automotive Engineers; an American society recognised world-wide for establishing vehicle standards and recommended practices.

Many different forms of drawings are used by engineers. These vary from simple freehand sketches to a variety of other types of drawing. List below different types of engineering drawings:

....

....

....

....

....

....

....

....

....

....

....

....

....

COMMUNICATING TECHNICAL INFORMATION

Communication of technical information can be achieved in a number of ways. List some of the methods by which information, in some standardised form, is made available in engineering and in the operation of a modern garage:

....

....

....

....

....

....

....

Give examples of the use of microfiche and visual display units in a modern garage.

MICROFICHE

....

....

VDU

....

....

....

What are the advantages of using microfilm or CD-ROM?

....

....

....

Rack and pinion steering gear

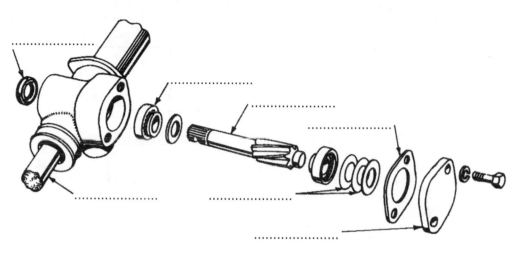

This type of drawing is called ....

Identify the arrowed parts on the drawing above and on the view of the same assembly below:

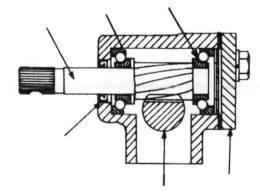

This type of drawing is called ....

207

TECHNICAL DRAWING

Orthographic Projection

Orthographic projection is a method used to present the various faces of an object when viewed squarely. For relatively simple objects, three views are sufficient to fully describe and give dimensions.

The three views are:

....

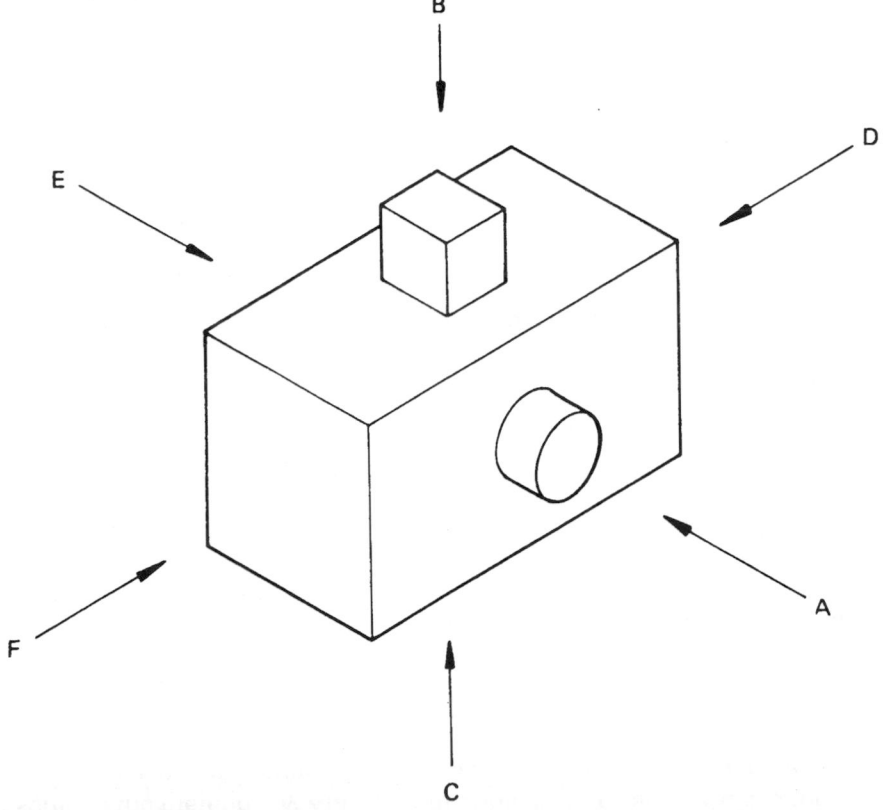

In which direction must the object be viewed to produce the views shown opposite, taking 'A' as the FRONT VIEW. Put the appropriate letter under the view.

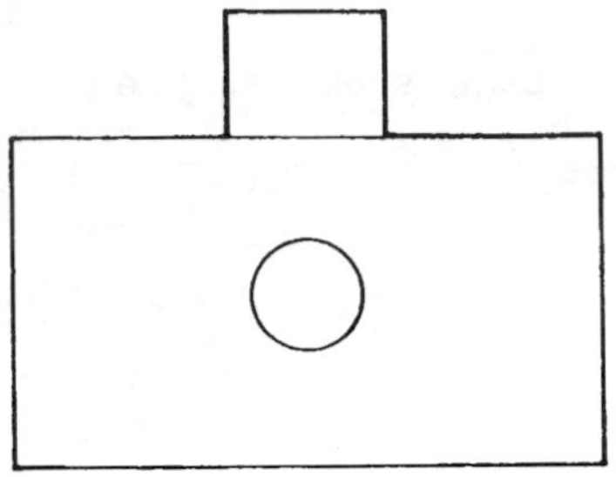

....A...

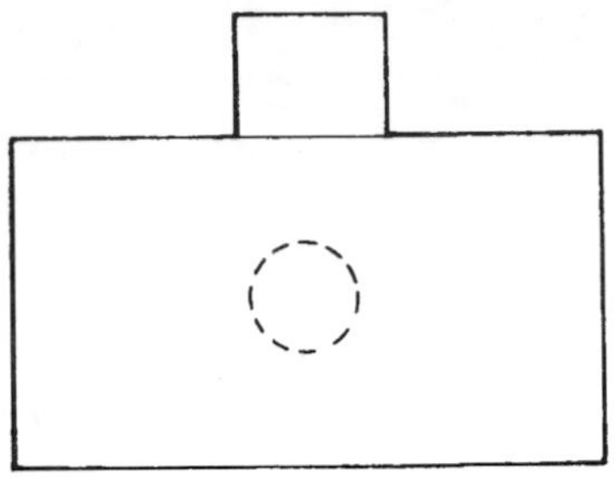

....

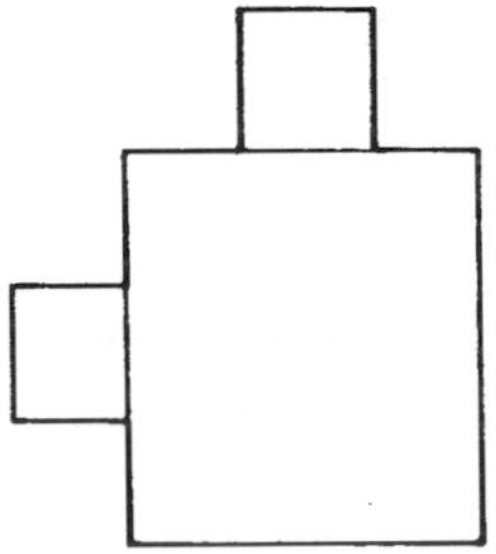

....

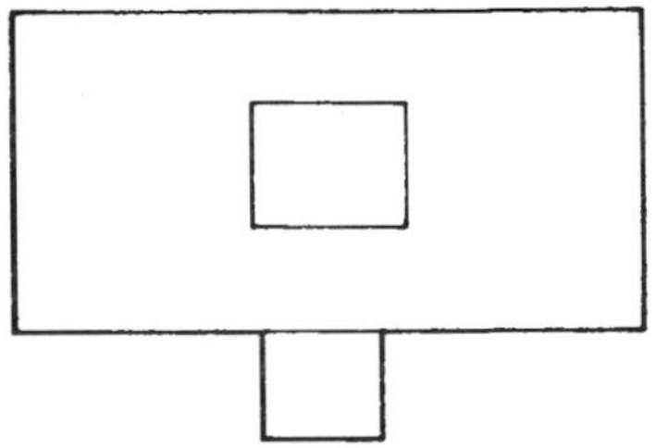

....

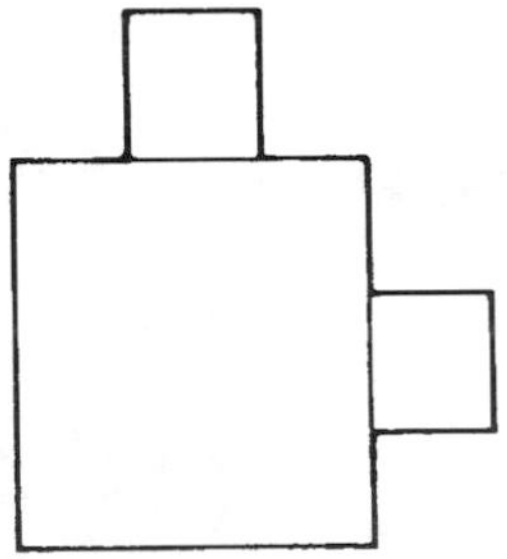

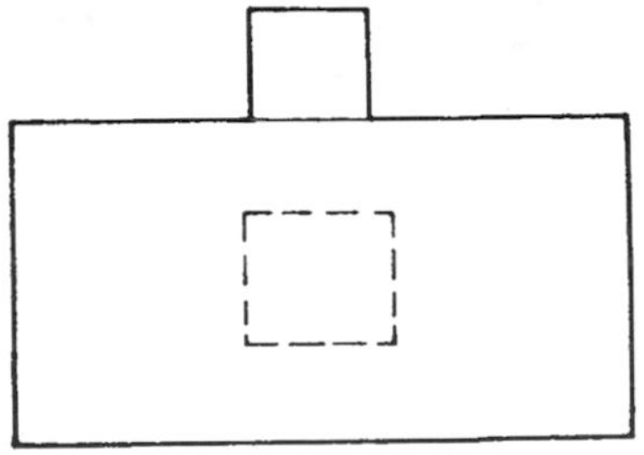

Pictorial Projection

A pictorial view of an object gives a three-dimensional impression where its height, width and depth are shown simultaneously.

Two methods of representing an object pictorially are by:

1. .... projection

and

2. .... projection

Features of isometric projection

1. The lines receding from the horizontal are drawn at .... to the horizontal.

2. Circles are drawn as ....

Features of oblique projection

1. The lines receding from the horizontal are drawn at .... to the horizontal.

2. The front face is drawn as a .... face.

3. The lines receding from the horizontal are drawn at .... full size.

Sketch two 2 cm cubes to illustrate isometric and oblique projection at the top of this page.

Name the method of projection under each of the pictorial drawings shown opposite.

Why are the receding lines reduced in length in oblique projection?

....

....

....

Isometric   *Oblique*

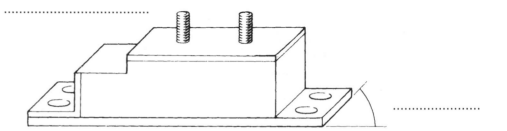

....

....

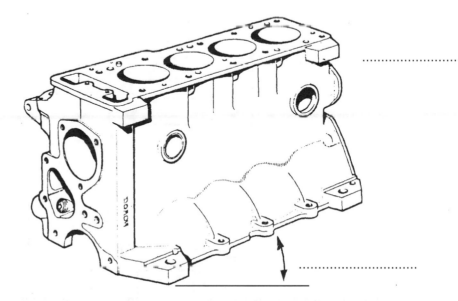

....

209

First Angle and Third Angle Projection

Two systems of orthographic projection are *first angle* and *third angle*. The difference between the two systems is the relative positioning of the three views on the drawing.

Consider the motor-vehicle battery shown below.

Name the angle of projection used in each 'three view' drawing below:

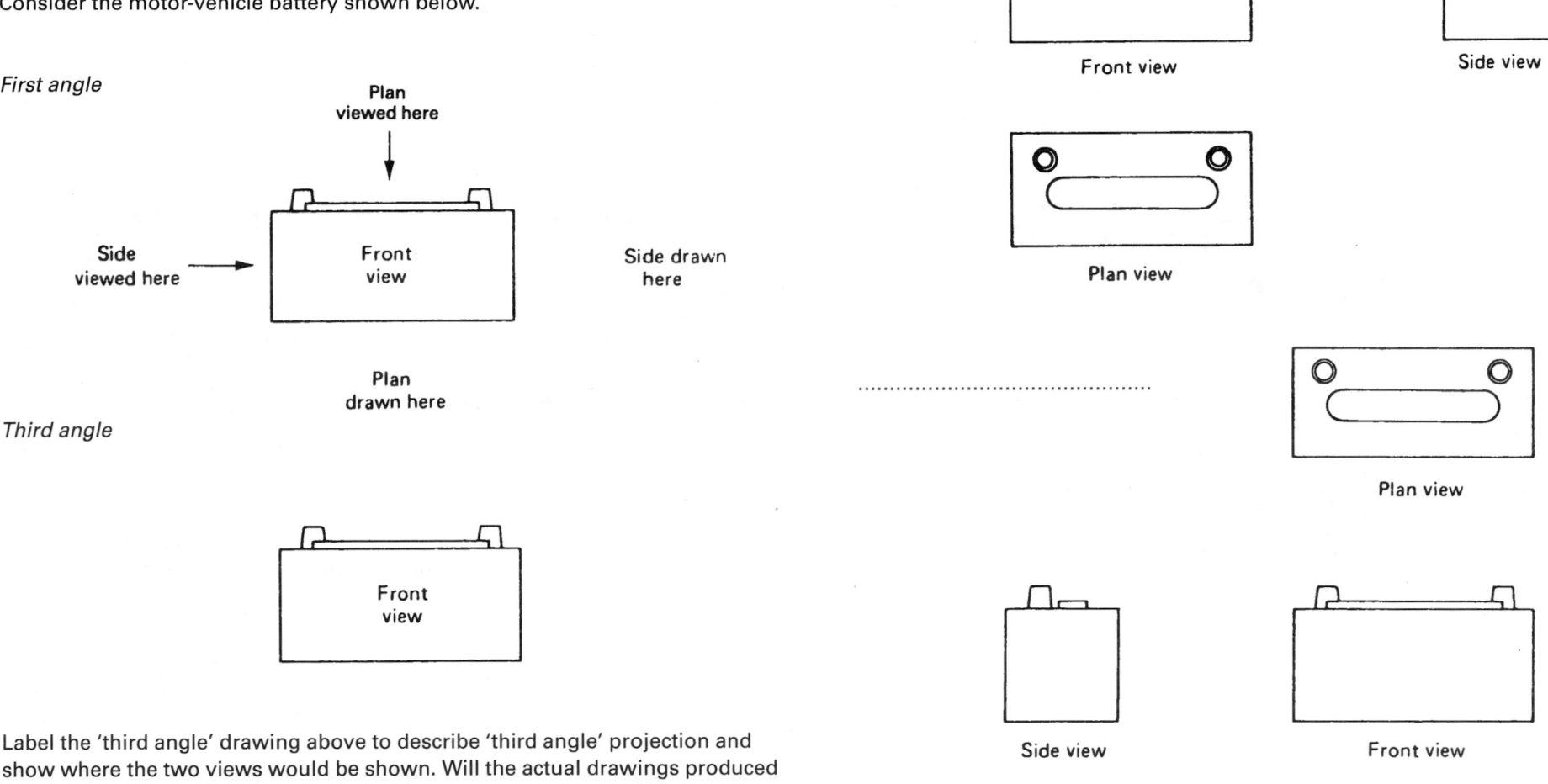

First angle

Plan
viewed here

Side
viewed here

Front
view

Side drawn
here

Plan
drawn here

Third angle

Front
view

Front view

Side view

Plan view

....

Plan view

Side view

Front view

Label the 'third angle' drawing above to describe 'third angle' projection and show where the two views would be shown. Will the actual drawings produced for first angle be different from the drawings for third angle projection? ....

....

....

....

The arrow on the pictorial drawings of the mounting brackets shown represents the front view. Study the views opposite and complete the table to indicate the correct front, side and plan views for the different brackets.

A

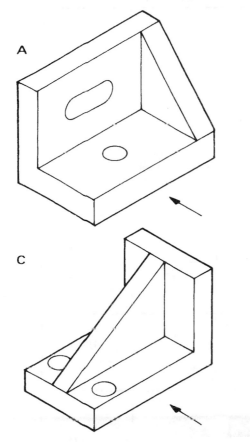

B

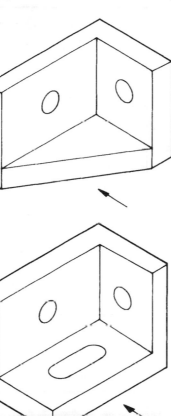

C

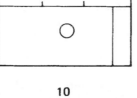

D

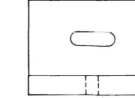

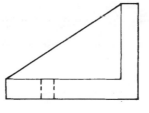

1

2

3

4

5

6

7

8

9

10

11

12

Pictorial	Front view	Side view	Plan view
A			
B			
C			
D			

These views could be either first angle or third angle depending upon their relative positions on the actual three view drawing.

211

Conventional Representation

To avoid wasting time on engineering drawings common features are often simplified, that is, they are shown by conventional symbols.

Complete the table to show the convention for the subjects shown:

TITLE	SUBJECT	CONVENTION
external screw threads		
splined shafts		
semi-elliptic leaf spring with eyes		
square on shaft		
bearings		

Information on Drawings

List typical examples of BASIC and ADDITIONAL information to be found on drawings.

Basic information

....

....

....

....

....

Additional information

....

....

....

....

....

Draw below the PROJECTION SYMBOLS for the two systems of projection.

Projection   **Symbol**

First angle

Third angle

What do the above symbols indicate?

....

....

....

....

GENERAL RULES

Types of lines

The types of lines used in engineering drawing are shown below. Complete the description of the lines by giving examples of their applications in drawing.

_____ Thick continuous

....

_____ Thin continuous

....

....

– – – – – – – – – Thin short dashes

....

— · — · — · Thin chain

....

━ · ━ · ━ · Thick chain

....

∿∿∿∿∿∿∿ Thin wavy

....

....

To produce the correct type of line to suit the application, not only requires some expertise but well-sharpened pencils of the correct grade. Give examples of the grades of pencils used for:

1. Visible outlines ....

2. Projection or centre lines ....

ABBREVIATIONS

By reference to BS 7308, complete the list at (A) below to give the recognised abbreviations for the terms stated and complete the list at (B) to give the terms for the abbreviations stated.

(A)	(B)
Drawing	SCR ....
Machined	SK ....
Hydraulic	SPEC ....
Millimetre	STD ....
Minute	U'CUT ....
Hexagon	PCD ....
Figure	NO ....
Centre line	CRS ....
Assembly	LG ....
Across flats	CHAM ....
Cylinder	SPH ....
Diameter	MATL ....
Galvanised	C'BORE ....
British Standard	NTS ....
Countersunk	RH ....
Centres	Int ....
Left hand	Max ....
Minimum	Sq ....

....

....

....

....

DIMENSIONING (MAIN RULES)

The simple drawing below is dimensioned in accordance with recognised standards.

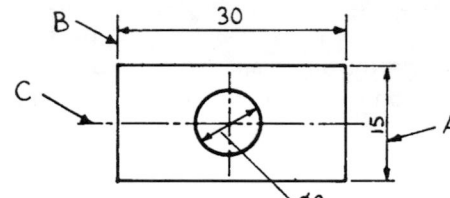

Line A is a ....

Line B is a ....

Line C is a ....

List the main rules for dimensioning a drawing:

....

....

....

....

....

....

....

....

....

....

....

The symbol φ represents ....

State the units used for linear dimensions on drawings:

....

Dimensioning from a Datum

A datum is a line, point or a face from which each dimension is measured, it is used as a base for a number of dimensions. A machined surface is often used as the datum.

Measure and dimension the drawing below taking (a) as the datum.

What is the reason for dimensioning from a datum?

....

....

....

....

BLOCK AND LINE DIAGRAMS

Block and line diagrams are used to illustrate, in a simplified form, component construction, layout of relative components or layout of certain systems (for example, hydraulic systems).

Name the type of diagram and the component or system illustrated below:

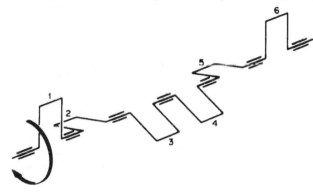

Diagram type .... Component ....

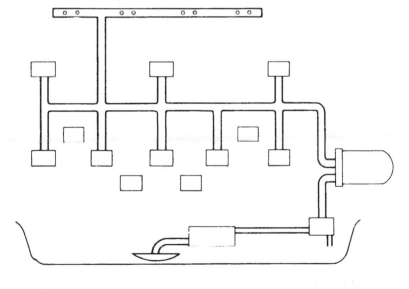

Diagram type .... Component ....

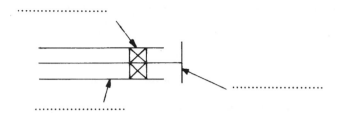

The simple line diagram above represents a ....

Label the diagram.

Illustrate below the layout of a petrol engine fuel system by using a block diagram.

The drawings on the left show a piston assembly and a valve. Make simple line diagrams on the right to illustrate the same.

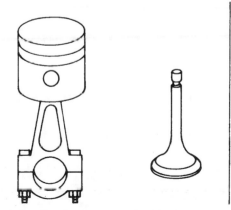

215

Assembly Drawings

This type of drawing shows two or more components assembled together. Usually dimensions are not shown but a list of parts is included.

State the purpose of the assembly drawing shown on the left:

....

....

....

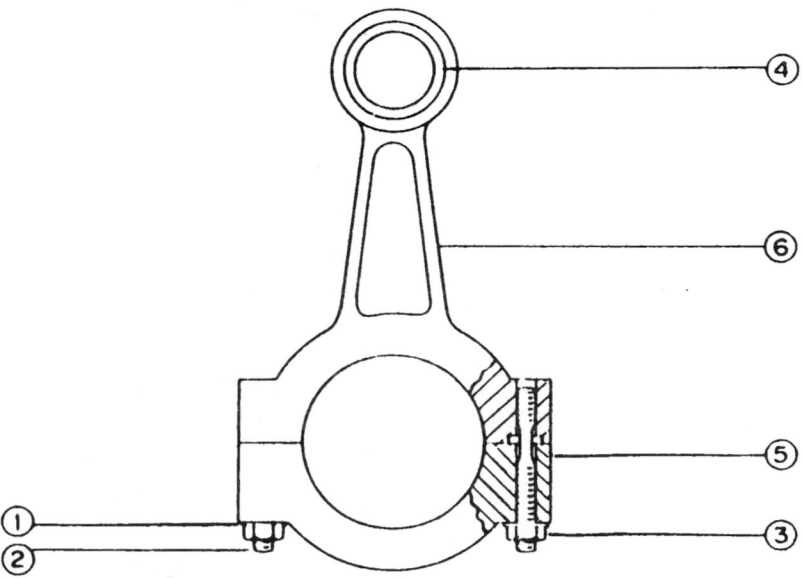

6	Rod	31127/6	Drop forged steel	1
5	Cap	31127/5	Drop forged steel	1
4	Bush	31127/4	Phosphor bronze	1
3	Nut	31127/3	Low carbon steel	2
2	Stud	31127/2	Low carbon steel	2
1	Washer	31127/1	Low carbon steel	2
Item	**Description**	**Drg. No.**	**Material**	**No. off**

Shown below is a typical bush used in motor-vehicle construction.

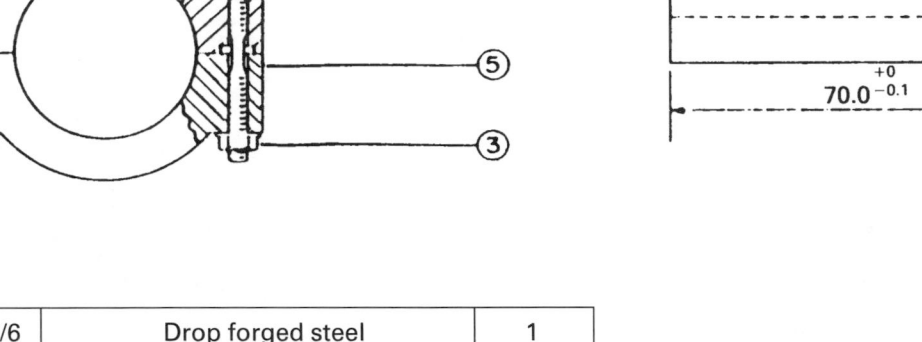

All dimensions in mm
Material – Phosphor bronze
Standard of finish – as turned

Name and state the purpose of the type of drawing shown above.

....

....

....

....

....

The type of drawing shown below is often referred to as a LOCATION DRAWING and is used frequently in repair manuals. State the role of the location drawing:

....

....

....

The view below shows a piston assembly in section.

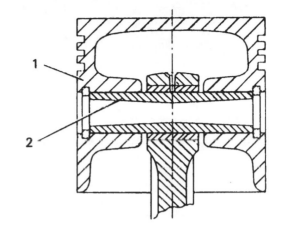

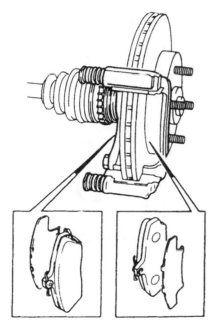

SECTIONAL VIEWS

To get a true indication of the actual cross-sectional shape of an object, it is often necessary to produce a 'sectional view' – as shown at right.

State two uses of a sectional view to a motor engineer:

1. ....

2. ....

   ....

What parts of the sectional view are cross-hatched?

....

....

....

Why are the cross-hatch angles different on parts 1 and 2?

....

At what angle to the horizontal do the cross-hatch lines lie?

....

To improve the clarity of a sectional view, certain parts when lying on a longitudinal cutting plane are not cross-hatched. These include such items as:

1. *Solid shafts and rods* ....

2. ....

3. ....

4. ....

5. ....

217

OPERATIONS SCHEDULES

An operations schedule outlines, in a logical sequence, the procedure to be followed when manufacturing a component or when assembling and dismantling a component. Very often the instructions within the schedule refer to a drawing or photograph to aid communication.

Draw up operations schedules for:

(1) the manufacture of the chassis plate shown in detail at (a), and

(2) the removal and replacement of the thermostat shown at (b).

(b)

(a)

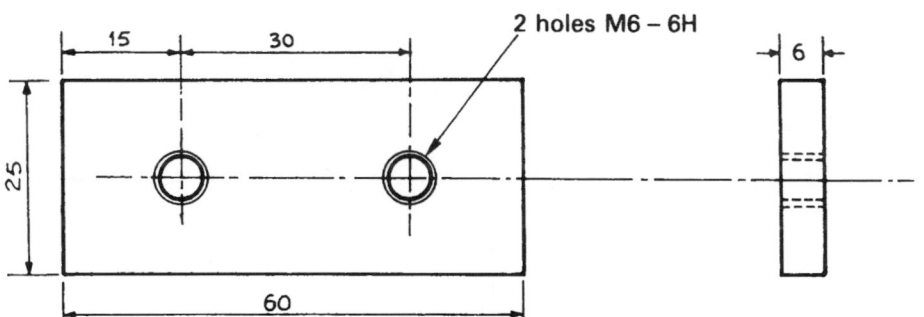

2 holes M6 – 6H

15   30   6

25

60

DATA INVOLVING SIMPLE MEASUREMENT

Name and state the purpose of the gauges shown below:

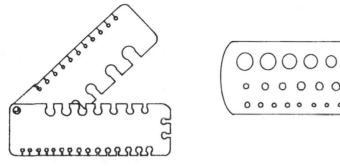

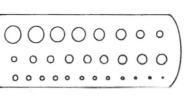

....

....

....

....

....

....

Use the tables in the workshop or stores to complete the table below giving twist drill sizes for the bolts indicated.

Clearance Hole Size (mm)								
Thread Diameter (mm)	2	3	4	5	6	12	16	20

Graphs

A graph is a means of communicating information; it gives a pictorial illustration of the relationship between two variables, e.g. Engine speed and Power, or Time and Distance.

The graph below shows the POWER and TORQUE curves for a modern car engine. Use the graph to interpret:

(a) Maximum engine power and speed at which this occurs

....

(b) Maximum engine torque and speed at which this occurs

....

(c) Torque at maximum power ....

(d) Power at maximum torque ....

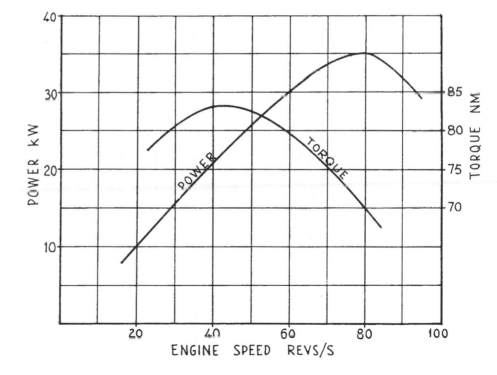

The graphical means by which information is conveyed depends very often on the subject and the type of relationship involved.

Bar and Pie Charts

An alternative method of communicating information in a pictorial way is by means of BAR or PIE charts. Construct a BAR chart opposite to show car sales figures over a six-month period using the following data.

Month	1	2	3	4	5	6
Sales	£120 000	£150 000	£100 000	£140 000	£170 000	£200 000

The PIE chart opposite shows the floor area occupied by the various departments of a garage business. Given the following data on floor area for the business, label the sections on the pie chart to illustrate the proportion of floor area occupied by each department.

Calculate the percentage area for each section of the pie chart.

WORKSHOP   –   2000 m²

CAR SALES   –   1500 m²

FORECOURT   –   1000 m²

PARTS DEPT   –   500 m²

BAR CHART

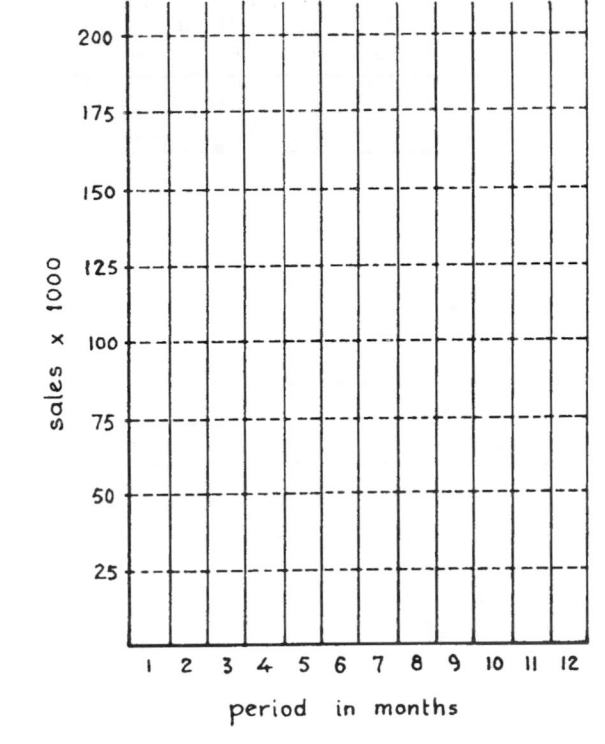

PIE CHART

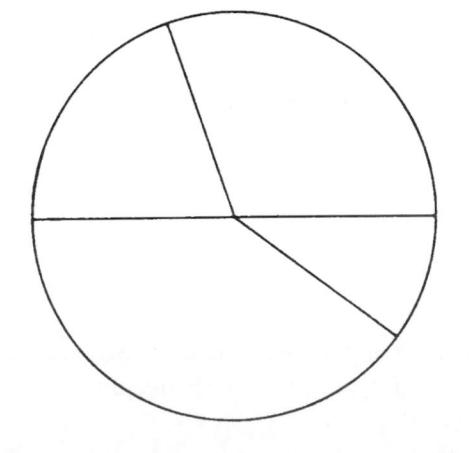

220

COLOUR CODING

Colour coding of components, equipment, piping, wiring etc. serves two purposes:

1. *It assists in identification when selecting and using, or when assembling or dismantling components or systems.*

2. ....
....
....
....
....

	Colour convention	
	Pipes	Cylinders
Oxygen		
Acetylene		
Compressed air		
Natural gas		
Butane		
Propane		

In accordance with current colour-coding conventions, complete the table above to indicate the colours for the items given.

Examine a modern vehicle and state the colour coding for lighting cables and fuses.

Vehicle make .... Model ....

Lighting cables

....
....
....
....

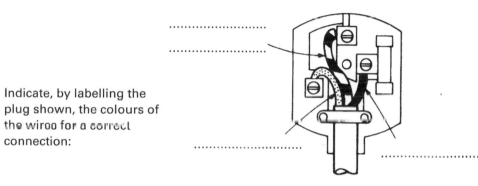

Indicate, by labelling the plug shown, the colours of the wires for a correct connection:

Fuses

....
....
....
....

Resistors in electrical circuits are colour coded to indicate their value. Examine a number of resistors. Show the colour coding on the resistors shown below for resistors of value: 68 kΩ   220 kΩ

Chapter 9

Bench Fitting

BENCH FITTING – METHODS USED

All technicians at some time in their work will be required to make (fabricate), modify or repair motor-vehicle components. To do this kind of work the skilled technician must have a basic knowledge of bench-fitting skills. These skills can be divided into specific categories and they form the basis of this chapter.

Examine the titles on the following pages up to the 'Care and Maintenance of Bench-fitting Tools', and list below the various bench-fitting operations.

1. ....
2. ....
3. ....
4. ....
5. ....

6. ....
7. ....
8. ....
9. ....
10. ....

MARKING OUT AND MEASUREMENT

When a simple component is to be made, lines are reproduced on the material's surface from an engineering drawing. This is known as marking out.

The tool used for marking straight lines on a component is called ....

State the purpose of the following with regard to marking out:

Dividers ....

Oddleg calipers ....
....

Bevel protractor ....
....

Combination set ....
....

Trammels ....
....

TOOLS USED

Identify the tools associated with marking out:

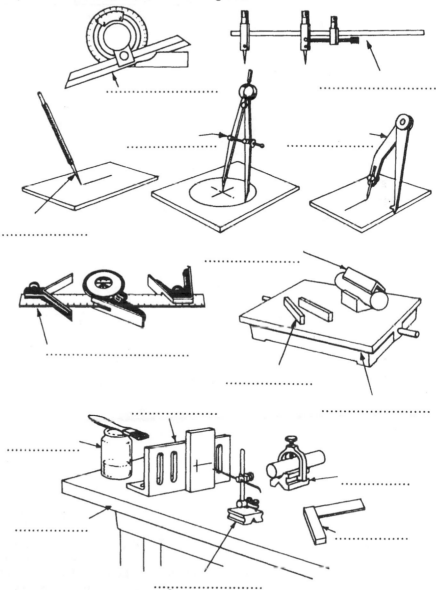

State a typical purpose of the following with regard to marking out:

Set square ....

....

Vee blocks and clamps ....

....

Surface table or surface plate ....

....

Angle plate and surface gauge ....

....

Parallels and parallel blocks ....

....

What is a template?

....

....

Why would it be useful to use a template to produce the following?

(a) 10 brackets as shown below.

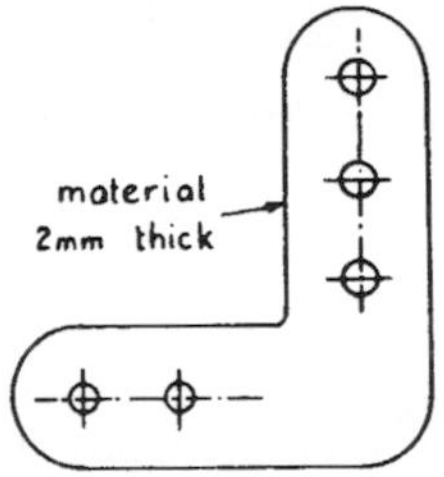

material
2mm thick

....

....

....

(b) A replacement piece for a lower wing.

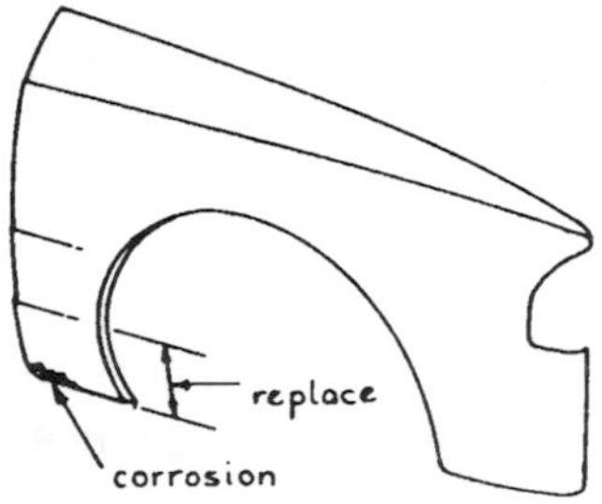

replace

corrosion

....

....

....

To enable the marking out on material such as mild steel to be seen clearly, the surface may be coated with some colouring. This may be:

1. ....   or   2. ....

The circles shown on the sketch below are centre-punched.

What is the reason for this?

....

....

What is the main use of the centre punch?

....

The component below has been marked out prior to cutting out and making.

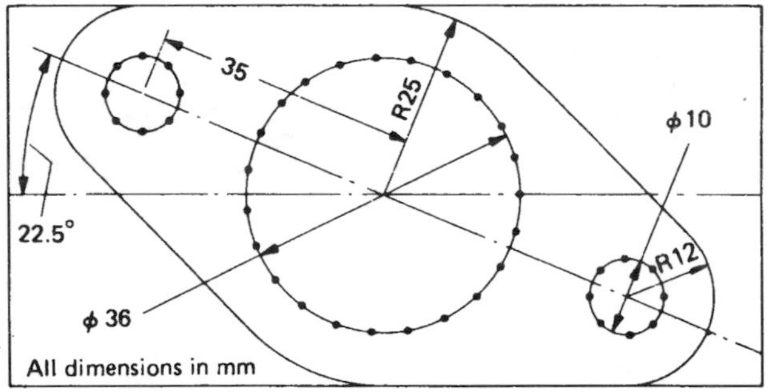

35   R25   φ10
22.5°   R12
φ36
All dimensions in mm

List the stages of marking out:

1.	
2.	
3.	
4.	
5.	
6.	
7.	
8.	

Identify these tools associated with measuring and gauging:

....

....

....

....

....

....

....

....

....

Which gauge would be particularly suitable for measuring wear in cylinder bores?

....

Note: These tools are covered in more detail in Chapter 7.

SAWING

The hacksaw is the most common type of saw used for cutting metals. The frame may be as shown below and may be adjustable or non-adjustable. Name the main parts of the hacksaws shown.

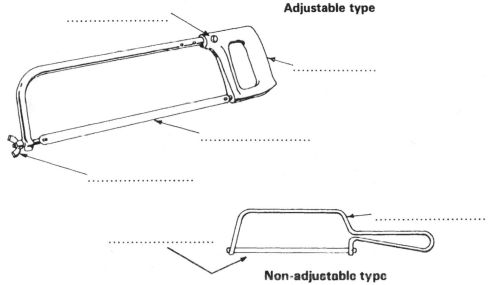

Adjustable type

Non-adjustable type

Hacksaw blades are made from heat treated high carbon steel and are classified according to their length and number of teeth per 25 mm.

The number of teeth (pitch) per 25 mm can be ....

The choice of blade depends upon the shape and type of material being cut.

Blades with teeth of 'fine pitch' are used for cutting

....

Saw blades with teeth of 'coarse pitch' are used for cutting

....

The individual teeth on a hacksaw blade form a pointed 'wedge' which digs into the metal and produces a 'shearing' action. The teeth on the blade are offset from one another. This offsetting is called the ....
of the teeth.

Examine two types of hacksaw blade. How do they differ with regard to their shape or 'set' of teeth?

....

....

Complete the sketch below to show a hacksaw blade in the sawcut. Indicate clearly the set on the teeth and state, alongside the drawing, the purpose of the set.

....

....

....

....

....

Which way round should the blade be fitted into a hacksaw frame?

....

On which stroke does the cutting action occur?

....

Two types of hacksaw blade are available, one is considered flexible and the other non-flexible. How do these blades differ in use?

....

....

....

....

Give three causes of saw blade breakage:

1. ....

2. ....

3. ....

Describe the procedure for sawing off a 300 mm length from a strip of metal whose cross-section is 25 mm × 10 mm:

....

....

....

....

....

....

....

....

When cutting thin sheet it is good practice to have at least 3 teeth in contact with the metal.

Show the correct position of the blade for sawing.

(i) *Tubing*   (ii) *Thin sheet metal*

....

....

CUTTING AND PUNCHING

Hammers

The hammer is a very common tool in all trades. On a motor vehicle, different types are used for special jobs. Name the hammers shown and give an example of their use. The ball and cross-pein hammers are available in different (sizes) weights.

....
....

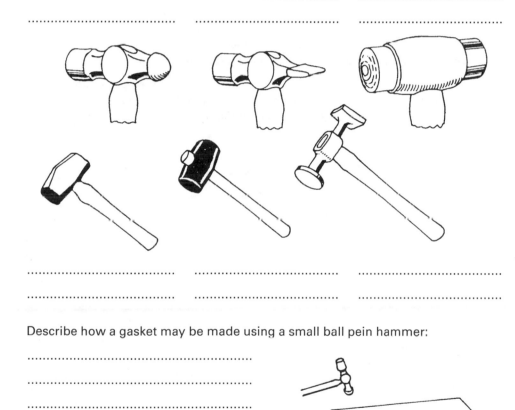

....
....

Describe how a gasket may be made using a small ball pein hammer:

....
....
....
....
....
....

Shears for Cutting Metal

Tin snips are commonly used for cutting thin sheet steel. The snips may be flat or curved-nosed. They are used very like a pair of scissors.
For cutting large sheets of metal (or metal too thick for snips) a guillotine or bench shears may be used.
Observe a demonstration cut with tin snips and say what happens to the narrow waste side of the metal.

....

Identify the items shown below:

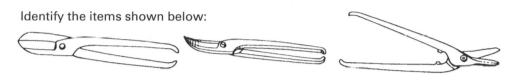

....

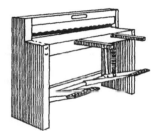 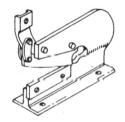

....

Wire Cutters

Below are shown two types of wire cutters. Name each type and describe its use:

1. ....   1. ....

....

2. ....

2. ....

CHISELS

The cutting action of a chisel works on the principle of forcing a wedge into the material to shear off any unwanted material.

The drawing below shows the point of a chisel during cutting. Hammer blows cause the pointed 'wedge' to 'shear' through the metal. The depth of cut is maintained by holding the chisel at the correct angle (angle of inclination). Three other important factors that affect the efficiency of the cutting action are:

1. .... angle,   2. .... angle

and   3. .... angle of the chisel.

Label the drawing.

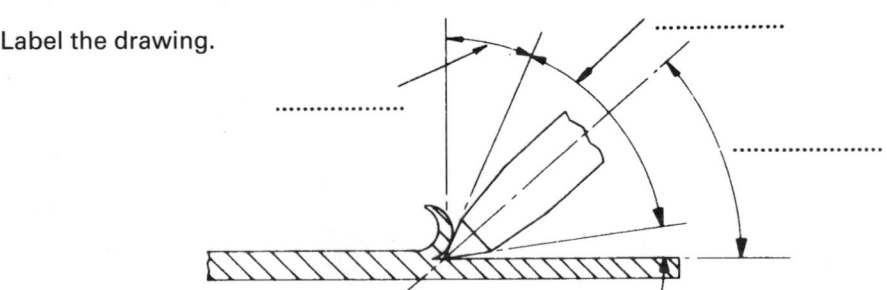

Chisels with point angles in the region of 55–65° are used for cutting relatively hard materials and the tool point is quite strong. For cutting softer materials the point angle is reduced.

The chisel points suitable for working the materials named below are shown. State the approximate point angle of each chisel.

Low carbon steel   *Aluminium*

The common engineering chisel is often called a COLD CHISEL. They are usually made from high carbon steel which is hardened only at the pointed end. Why is the head left soft?

....

....

Examine chisels of the types named below, and in the spaces provided make sketches of each type and give examples of use.

Flat chisel   *Cross-cut chisel*

....

....

Diamond point chisel   *Half round (round nose) chisel*

....

....

When chiselling and striking with the hammer, on what part of the work should the eyes be focused?

....

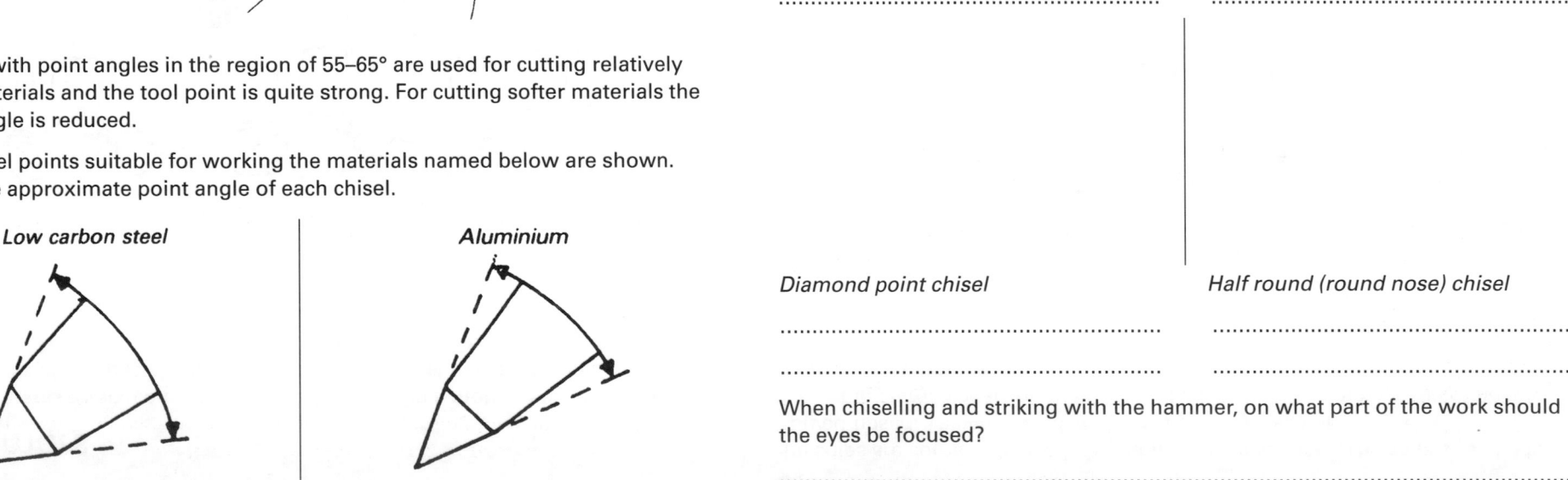

228

FILING

The cutting action of a file is similar to that of a chisel or hacksaw. Each tooth on a file is a tiny cutting blade. Files are classified according to: length, shape, grade of cut and type of cut.

Name the types of file shown below.

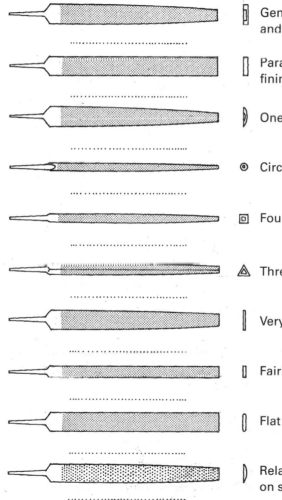

General utility file, cuts on both sides and edges.

....

Parallel in width, used for roughing or finishing. One safe edge

....

One side flat, one curved

....

Circular section, tapering

....

Four equal sides, 90° angles

....

Three cutting sides at 60° angles

....

Very thin file

....

Fairly slim file. One safe edge

....

Flat faces but radiused edges

....

Relatively large, separate teeth. Used on soft materials

....

As can be seen, there is a shape of file to suit any work situation.

The file teeth or angles of file teeth are shown below. Indicate the teeth angles in the second and third drawing.

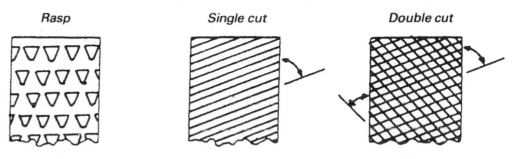

Files are graded according to the number of teeth per centimetre, that is, their roughness or smoothness. Name four grades of file between the rasp and dead smooth files and state typical uses.

File grade	Typical use
....	....
....	....
....	....
....	....
....	....
....	....
....	....
....	....

State four hints for obtaining maximum life from a file:

1. ....

2. ....

3. ....

4. ....

DRILLING

The most common type of drill is the twist drill. It is supplied in three lengths: jobbers' series, these are normal length drills; long series; and stub (short) drills. Metric drill sizes range from 1 mm diameter to 20 mm diameter, the smaller sizes being available in 0.1 mm steps.

Name the main parts of the drills shown, including the different types of shank:

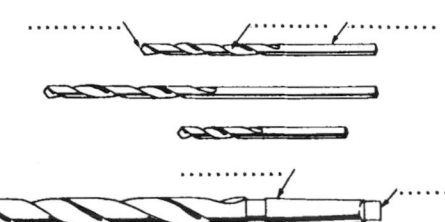

Other types of drills are shown below. Name and state suitable uses for them:

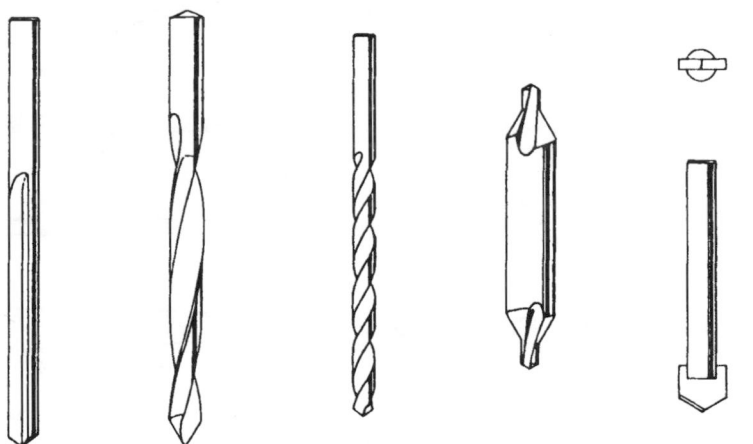

....

....

....

....

....

....

Twist drills remove metal by using two cutting edges rotating about a centre point.

Indicate the cutting edges and lands. Indicate included (point) angle. Show the direction of rotation.

The point angle necessary for general-purpose work is ....

State the purpose of the *land* on a twist drill:

....

....

When a drill is sharpened by grinding, it is essential to keep the cutting edges (lips) the same length and at the same angle to the drill axis. State how each drill below has been incorrectly sharpened.

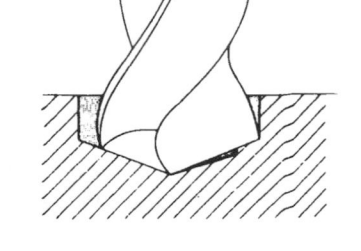

....

How is the hole affected by the faults shown above?

....

The above are causes of inaccurate drilling.
State TWO other causes:

....

....

A soluble oil (mixed with water) should be used when machine shop drilling low carbon steel, aluminium, copper and phosphor bronze. Cast iron, brass and plastics can be drilled dry. When using a hand-drill on a vehicle all materials will be drilled dry, so care must be taken to avoid overheating.

Name the types of drilling machines shown below:

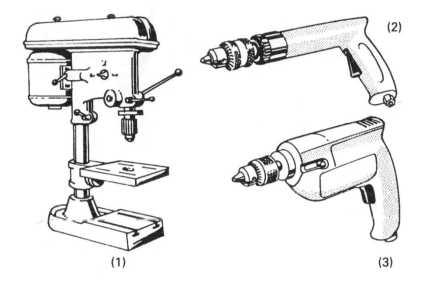

(1)   (3)

State how the drilling speed is varied on

Drill (1) ....

....

Drill (3) ....

....

What are common operating voltages for the electric drills shown?

....

Drill Speeds and Feeds

The *spindle speed* of a drill is the rotational speed at which the drill turns.

It is expressed in ....

The *linear cutting speed* of a drill is the speed at which the cutting edges pass

over the work. it is expressed in ....

What is meant by the *feed* on a drilling machine?

....

....

As a general guide, the feed increases as the drill size increases.

The spindle speed of a drill is set according to:

1. ....

2. ....

State the linear cutting speed (using HSS drill) for:

1. Tool steel .... 2. Mild steel .... 3. Brass ....

NOTE: High-speed steel (HSS), used for most twist drills, is an alloy steel in which the main alloying element is tungsten. If plain carbon steel is used for a twist drill, the cutting speed must be considerably reduced.

It can be seen from the examples chosen that, when cutting soft material, the cutting speed is higher than that used for hard material. It is normal practice to refer to tables to obtain the correct spindle speed for a particular material and size of hole required.

Table of approximate drilling speeds (rev/min)

Drill diam.	Type of metal			
	Steel	Cast iron	Brass	Aluminium
4	1200	1900	6000	7200
6	800	1270	4000	4800
8	600	900	3000	3600
10	480	765	2400	2800
12	400	640	2000	2400

When drilling a hole what precautions could be taken to ensure accuracy of the operation?

Show on the hole where 'cut back' grooves should be positioned:

....
....
....
....
....
....
....
....
....

When cutting sheet metal, what precautions should be made to prevent damage to the work when using a conventional drill?

....
....
....
....

Name the alternative types of drill that can be used to cut sheet metal:

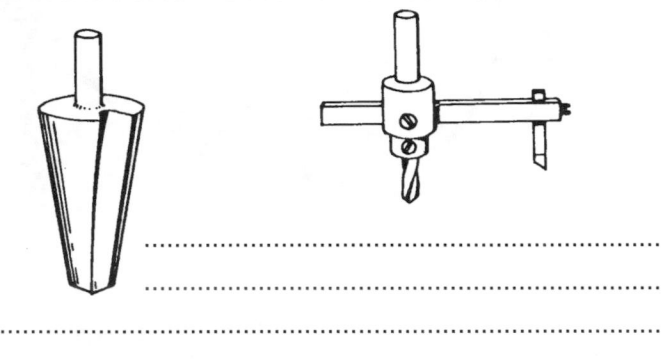

....
....
....
....

Other processes that can be carried out on a drilling machine are shown below. Describe the function of each operation.

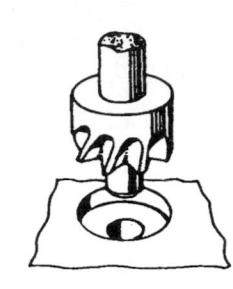 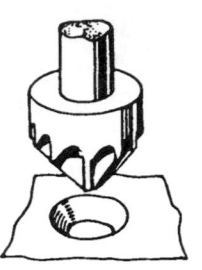

spot facing	counter boring	counter sinking

spot facing
....
....
....
....

counter boring
....
....
....
....

counter sinking
....
....
....
....

PUNCHES

Name the types of punches shown below:

Where would each type be used?

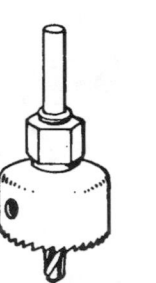

....
....
....
....
....
....
....
....
....
....
....
....

SCREW THREAD CUTTING

A screw thread is produced with the aid of a tap or die.

Taps

The tap is an accurately made thread with a cutting edge formed on the thread. There are three types of tap. These are shown below. Name each type and identify the types of hand wrenches that can be used to operate the tap.

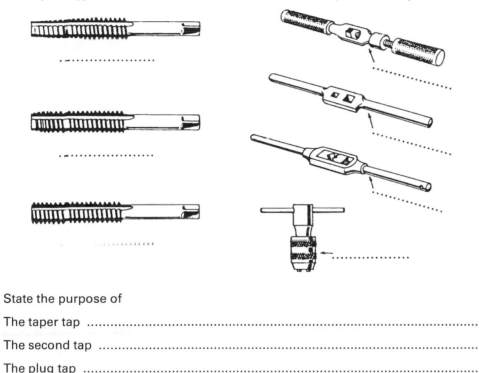

State the purpose of

The taper tap ....

The second tap ....

The plug tap ....

....

When commencing to tap a drilled hole of the correct size, the correct procedure is to assemble taper tap and rotate with a downward pressure until tap starts cutting. Check for squareness, pull square while turning until correct. Continue tapping without downward pressure – two forward rotations to one backward. Continue until tap cuts through hole. Remove and use second tap. A good thread is obtained using a cutting lubricant.

What precautions should be taken when tapping a blind hole?

....

....

....

What is the purpose of the item shown below?

....

....

....

....

....

....

Dies

A die is a form of nut accurately made with a cutting edge formed on the thread. Why is the die either split or made up of two pieces?

....

....

Name the types of stocks and dies shown:

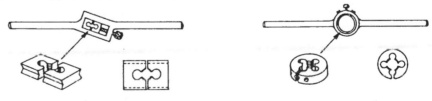

....

When commencing to cut a thread on a bar using a die, the correct procedure is to tighten button die so that it is in its open position. Squarely place stock and die on bar and with a downward pressure rotate clockwise. When die starts cutting check for squareness and realign. Cut using three forward rotations to one backward rotation.

233

ISO METRIC SCREW THREAD

Screw threads may be external (male) – for example, bolts, studs, screws – or internal (female) – for example, nuts and threaded holes.

The metric thread is now used very extensively in motor-vehicle engineering. A coarse or fine pitch is used to suit the application.

Sketch below the profile of a metric screw thread:

Complete the labelling below to show the main terms used in describing screw threads:

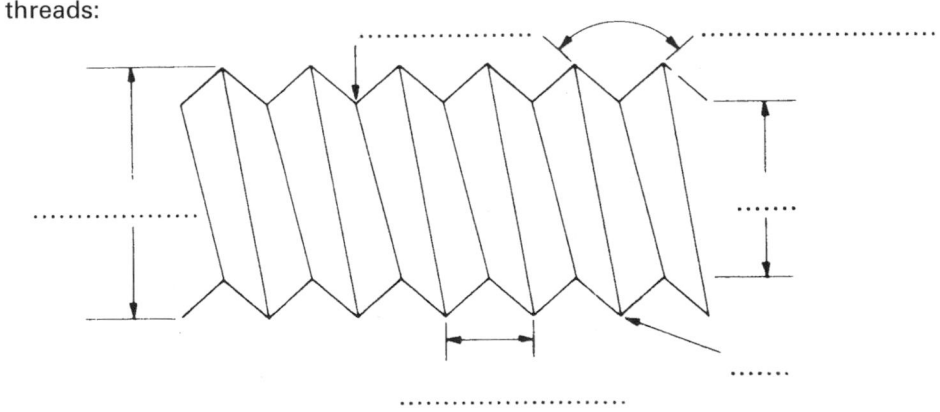

Thread designation

A thread designation such as M10 × 1.5 describes a metric thread, where

M = ...   10 = ...   1.5 = ...

ISO METRIC is an abbreviation for ....

Define the term *thread depth* ....

....

Define the term *lead* used in connection with screw threads.

....

....

Fine pitch and coarse pitch screw threads are shown below, label the drawings accordingly.

....

Complete these statements regarding coarse and fine threads such as those shown above.

(a) The thread having the greater pitch is the .... thread.

(b) The greater thread depth is on the .... thread.

(c) For one turn of the nut on the fine thread it will move a ....

    distance than would a nut on the .... thread.

    Its .... is less.

Why are course threads often used in soft alloys?

....

....

....

Why are fine threads very commonly used for most motor-vehicle applications?

....

....

TYPES OF SCREW THREAD

Although metric threads are normally used, there are still a large number of other screw thread forms found on motor vehicles.

The abbreviations for these are shown in the tables below.

Complete the table to show what the abbreviations represent:

BSW	British Standard Whitworth
BSF	
BSP	
BA	
UNC	
UNF	
ANF	
ANC	
APT	

The correct size of hole to be drilled for cutting a particular thread can be determined by referring to 'thread tables'.

Extract from ISO metric coarse thread table

Screw or bar diam.	Screw designation (diam. × pitoh)	Tapping drill diam.	Clearance drill diam.
4	M 4 × 0.70	3.30	4.8
6	M 6 × 1.00	5.00	7.0
8	M 8 × 1.25	6.80	9.0
10	M10 × 1.50	8.50	11.0
12	M12 × 1.75	10.20	13.0

Indicate each dimension on the sketch for a 10 M screw size:

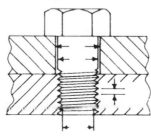

State a workshop method, other than using tables, to determine:

(a) The approximate drill size for cutting a thread ....

....

(b) The clearance hole diameter for a bolt to pass through ....

....

Name the items shown below and state how they would be used in the repair of vehicle components:

....

....

....

....

....

....

....

....

....

....

....

....

....

....

....

....

Describe how the last item shown above should be used in the situation shown below:

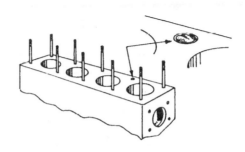

....

....

....

....

....

....

....

REAMING

If a very accurate hole size is required, then a reamer should be used to achieve this hole size. This is because an ordinary high speed drill may not drill with 100% accuracy. Name the type of reamers shown below:

....

....

....

....

....

....

Why are the parallel reamers 'slightly' tapered at the end?

....

Why do some reamers have left-hand spiral flutes?

....
....

When would the taper reamer be used?

....

What type of lubrication is required?

....
....

When using a reamer, a hole of suitable size should have first been drilled. Depending on drill size, this hole is 0.2 to 0.5 mm less than the required hole finish size. Complete the table below to show drill size prior to reaming:

Diameter of hole finished size (mm)	Drill size before reaming (mm)	Reaming allowance (mm)
4		0.20
8		0.20
12		0.25
16		0.50
20		0.50
30		0.50

What precautions should be observed when hand reaming a hole?

....
....
....
....

What is the advantage of using adjustable reamers?

....
....

To set up an adjustable reamer when the hole size is known, use a micrometer or ring gauge to adjust the reamer. The blades are in tapered channels fixed with a lock ring top and bottom. The blades move out when the top ring is slackened and the bottom ring moved up.

What is the purpose of pilot reamers?

....
....
....

GRINDING

Most vehicle repair shops will possess some or all of the types of grinding machines shown below.
Name the types shown.

State PERSONAL PRECAUTIONS to be observed when using abrasive wheels:

....
....
....
....

Before commencing any grinding operation, a visual inspection of the grinding wheel's and the machine's condition should be made.
List the important inspection points:

....
....
....
....
....
....

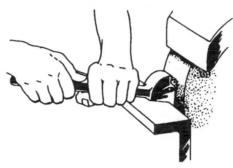

Dressing a wheel

Grinding speeds

Over-speeding is one cause of a grindstone wheel bursting.

Wheel bursting

This accounts for 12% of grindstone accidents, many of which are fatal.

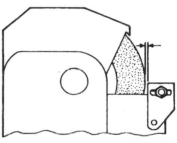

Recommended exposure of wheel

It is important that the speed of abrasive wheels is not exceeded. Doubling the speed increases the centrifugal stress four times. What are the regulations concerning speed?

....
....
....
....

What is the main cause of over-speeding?

....

WORK HOLDING
The Bench Vice

Two types of vice are commonly found in motor-vehicle workshops. These are shown below. They look very similar but one is adjustable (quick release) and the other is non-adjustable.

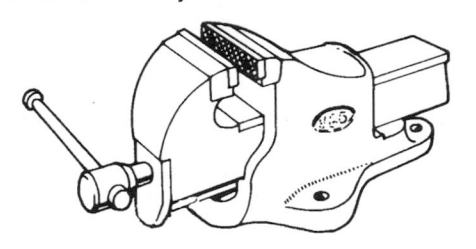

....

Both types can be supplied with a swivel base. The **'garage' vice** shown below has a swivel base, hardened steel anvils and offset jaws to hold long and wide work vertically.
State other features that the vice possesses:

....

....

....

....

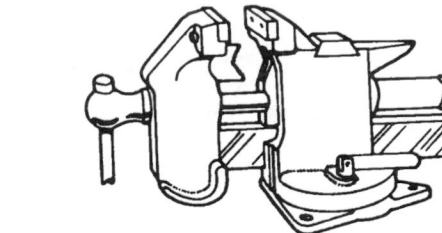

When machine drilling for light machine operations, a hand vice is commonly used while the drill is held in a chuck. Name the workholding clamps shown:

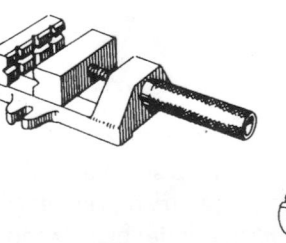  

Clamps

For holding small work, hand clamps are suitable. Name the clamps shown:

....

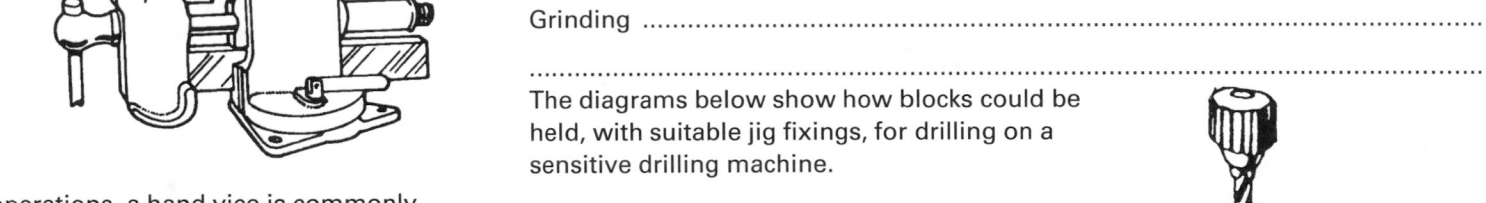

....

When drilling for absolute accuracy on a sensitive drilling machine, the material must be securely clamped. Name the jig clamps shown:

....

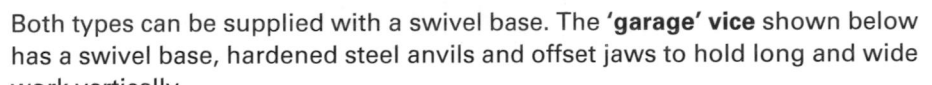

....

State the positions of the holding device when carrying out the following operations:

Cutting ....

Drilling ....

....

Grinding ....

The diagrams below show how blocks could be held, with suitable jig fixings, for drilling on a sensitive drilling machine.

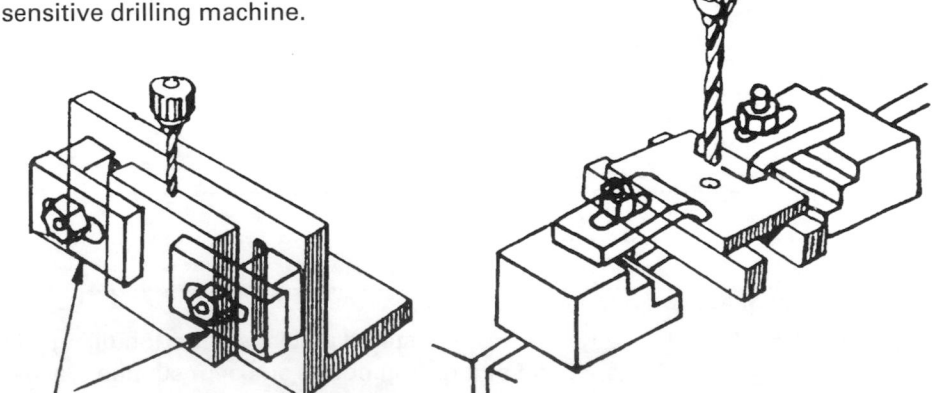

PIPE BENDING

For bending conduit and heavy copper pipe, bench-mounted bending machines are used.

Plumber's light gauge bender

Small-diameter pipes can be bent using a block of wood that will rest on the floor and has a hole of suitable size at the top end. The pipe is put through the hole and pressed at both sides until it is bent to the shape required. How can bends and kinks be prevented when bending in this manner?

....

....

....

....

Brake Pipe Bending

Two types of pipe or tubes which may require bending in a garage workshop are:

....

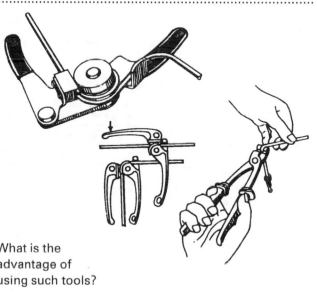

What is the advantage of using such tools?

....

....

Bend Allowance

If pipes are bent without allowing for the metal shortening on the inner corners, the finished pipe will be undersize and difficult to fit.

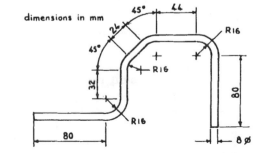

Indicate on the drawing the radii used for calculating the curves.
Calculate the length of pipe required to make the shape shown.

....

....

....

....

CARE AND MAINTENANCE OF BENCH-FITTING TOOLS

Select TWO tools from examples such as files, drills, reamers, hacksaw, chisel; examine them and complete the table below by considering their care and maintenance.

Selected tools		
Protection		
Checking for truth		
Handling		
Sharpening		
Selection		
Safety		

COMMON FORMS OF METAL SUPPLY

The metal stocked by a garage is mostly in the form of finished products, e.g. engines, gearboxes, exhausts and body panels. For basic repair it may stock some sheet metal for body fabrication repair, and various flat and round bars or other section tubes.

Name the standard shape of bar shown below:

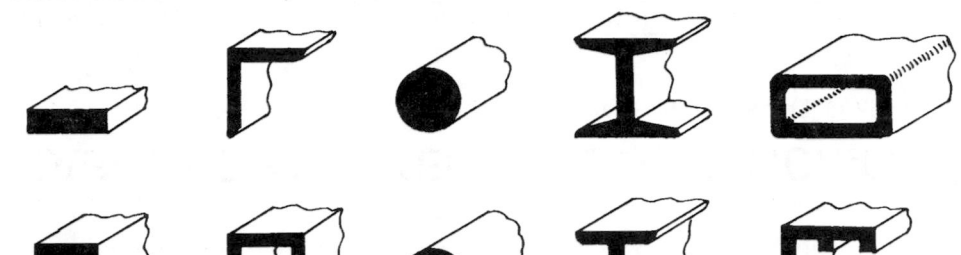

Colour coding

A store may stock many grades of metal which all look the same but have totally different properties and are required for different uses. In order to be able to identify these metals easily a colour code has been introduced and colours are painted on the different metals. Manufacturers of metals can provide specific details of these codes to enable users to select the correct grade for a particular job. The chart below gives examples.

	Ferrous metal			Non-ferrous metal		
	BS No.	(1995 No.)	Colour	BS No.	Material	Colour
Steels	220M07	EN1A	Green	2874	Brass	Brown/White
	070M20	EN3	White	369	Bronze	Purple/Red
	080M30	EN6	Black	1400	Lead bronze	Purple/Green
	401S45	EN52	Red/White Blue	3037	Monel	Purple/Orange
	431S29	EN57	Yellow	1474	Aluminium	Brown/White
	Cast iron		Grey	–	Copper	Self Colour

FERROUS AND NON-FERROUS METALS

Most of the main mechanical and structural components of a car are made from metal.

Metals may be split into two main groups: Ferrous and non-ferrous metals

A ferrous metal is ....

....

A non-ferrous metal is ....

....

....

Name typical motor-vehicle components made from the materials below:

Ferrous metals	Components	Non-ferrous metals	Components
Low-carbon (mild) steel		Aluminium alloy	
High-carbon steel		Copper	
Cast-iron		Copper-based alloys, brass and bronze	
Alloy steels		Lead-based alloys	
		Zinc-based alloys (diecasting)	

Ferrous Metals

Ferrous metals are defined by the amount of carbon contained in the metal:

....

....

....

State which of the following metals will have the percentage carbon content shown opposite:

Cast iron
Mild steel
High-carbon steel
Wrought iron
Medium-carbon steel

Material	% Carbon
	0.01
	0.25
	0.50
	1.20
	3.00

Alloy Steels

Alloy steels are used for most of the highly stressed components used in the modern car. What is meant by an alloy steel?

....

....

....

....

....

Most of the tools in a technician's tool box are made from alloy steel, for example, spanners, pliers, hammers, screwdrivers, chisels, hacksaw blades.

Examine a selection of spanners. From what alloy steel are they made?

Spanner	Type of alloy steel

Why are alloying elements such as those listed below added to steel?

....

....

The table shows how each of the elements can be used to improve the properties of steel.

Elements	Properties improved
Nickel	Toughness and ductility
Chromium	Hardness
Nickel and chromium	Hardness, heat and corrosion resistance
Chromium and vanadium	Electricity and fatigue resistance
Chromium and molybdenum	Less brittle at high temperatures
Tungsten	Hardness and strength at high temperatures
Manganese	Fatigue resistance
Cobalt	Increased hardness at high temperatures

List some typical steel motor-vehicle components and state the type of alloy steel from which they may be made.

Typical motor-vehicle components	Typical alloy steel material

Non-ferrous Metals

Most of the pure non-ferrous metals are not used separately but are alloyed with other materials when used to produce motor-vehicle components.

The reason for this is ....

....

....

State the main properties of the non-ferrous metals shown below:

Material	Colour	Main properties
Aluminium		
Copper		
Tin		
Lead		
Zinc		

Aluminium and its alloys

Pure aluminium is not commonly used in a vehicle because it is too ductile and malleable. But, when small amounts of other materials are added, alloys can be produced that are much stronger, harder, able to retain strength at high temperatures and corrosion resistant. Describe two types of aluminium alloy.

....

....

....

....

....

....

....

Give reasons why the following non-ferrous metals are considered very suitable materials for the following components:

Component	Material	Reason for choice
Piston Some cylinder heads	Aluminium	
Radiator core (or stack)	Copper	
Electrical cables	Copper	
Fuel pumps, carburettors	Zinc-based aluminium alloy	
Small plain bearings	Bronze	
Radiator header tanks	Brass	
Bearings (thin shell)	Aluminium, tin, copper, lead	

The main alloys of copper are brass and bronze.

Brass is an alloy of copper and ....

Bronze is an alloy of copper and ....

PLASTICS

Plastics are a large group of man-made materials. They may be formed into any required shape under the application of heat and pressure. There are two groups of plastics. State their properties and give a typical use.

1. Thermosetting

....

....

....

....

....

....

2. Thermoplastic

....

....

....

....

....

....

State whether the following materials are thermoplastic or thermosetting:

Material	Type	Material	Type
Celluloid		Bakelite	
Formica		Polythene	
Polystyrene		PVC	
PTFE (Teflon)		Nylon	
Terylene		Epoxy resins	

Listed are three common plastics used on vehicles.
Complete the table as required.

Material	Properties	Vehicle applications
Nylon	Strong Heat resistant Low coefficient of friction.	
PTFE	Self-lubricating.	
PVC	Good insulator. Can be coloured. Chemically resistant. Reasonable heat resistance.	

Some body panels and one-piece car bodies are manufactured by using thermosetting plastics (polyester resins) to reinforce glass fibre.
State advantages of this type of body over an 'all-steel' body:

1. ....

2. ....

3. ....

4. ....

CERAMICS

A ceramic material is one that has been produced through a heating process and forms a pot-like substance that is very hard and brittle.

One of the few ceramic substances used on a vehicle is for:

....

....

For adhesives, see Chapter 10.

STRESS

Engineering materials are designed to withstand the external forces and pressures placed upon them. The strength of a material is determined by its ability to resist loads without breaking.

The internal reaction that is set up in a material when force is applied is known as STRESS and the (often small) deformation that takes place when stress occurs is known as STRAIN.

Name the types of stress induced in the components shown at (a), (b), (c) and (d) by the application of the forces illustrated.

(a)

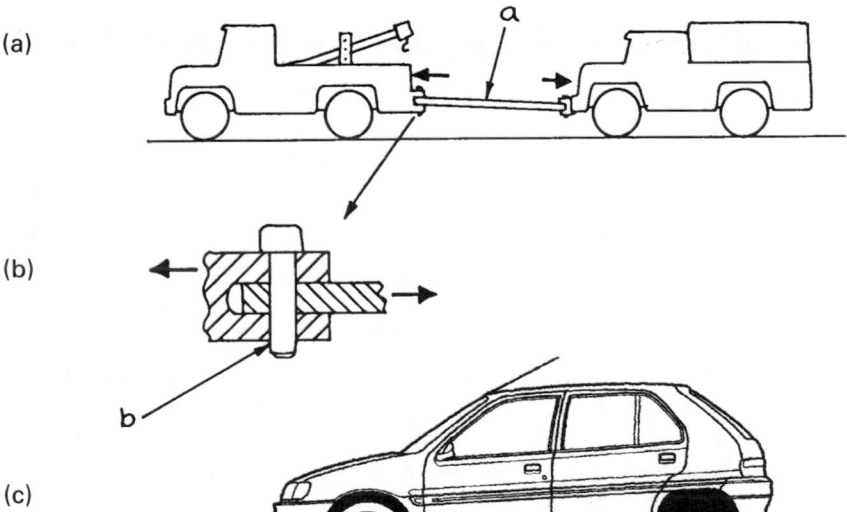

(b)

(c)

(d)

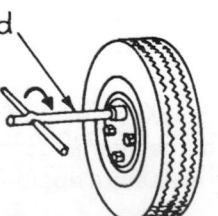

State the type of stress to which the following motor-vehicle components are mainly subjected:

Component	Types of stress
Cylinder head bolt	
Connecting rod	
Clutch disc rivets	
Handbrake rods	
Coil spring	
Cylinder head gasket	
Gudgeon pin	
Propeller shaft	

Stress is directly proportional to the force applied.

$$\text{Stress} = \underline{\hspace{5cm}}$$

Tension

When a tensile force is applied to a material, the material will stretch and will be in tension. As the force increases, the material will continue to stretch or elongate until it breaks.

The ease with which a material stretches, the amount of elongation before breakage and the force it can withstand before breakage, all depend on the type of material.

CHARACTERISTICS (OR PROPERTIES) OF MATERIALS

The selection of materials for a particular application is determined by their characteristics.
A material may possess one or more of these characteristics.
Define each characteristic and give an example of a motor-vehicle component which demonstrates this characteristic.

Characteristics	Description	Typical MV component
Hardness		
Strength		
Brittleness		
Toughness		
Ductility		
Malleability		
Elasticity		
Plasticity		
Softness		
Thermal conductivity		
Electrical conductivity		

MATERIAL SELECTION WITH REGARD TO ITS PROPERTIES

Low-carbon steel or mild steel

List the properties that make low carbon steel a very suitable material for car bodies:

....

....

....

....

....

Cast iron

List the properties that make iron a particularly suitable material from which to make such parts as engine cylinder heads and blocks, cylinder liners, piston rings, brake drums and clutch pressure-plates:

....

....

....

....

....

EFFECTS OF MATERIAL PROPERTIES ON ASSEMBLING AND DISMANTLING

All types of mechanical fastening should be tightened to their correct torque and this value should not be exceeded.

Bolts are graded according to their tensile strength. How can they be identfed?

....

....

....

What is the advantage of using bolts of high tensile strength?

....

....

....

It is possible for bolts to be made stronger although produced from the same material. Indicate the principle on the sketches below relative to machined and forged-rolled bolts:

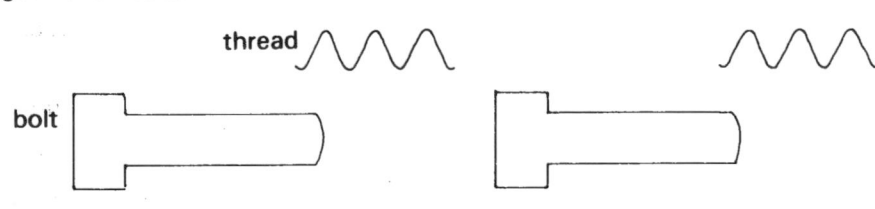

The material of a component must be known before it is tightened because different materials have different properties. For example, some may stretch easily, others may be very brittle.

State a type of material that is brittle ....

State two important considerations relative to assembling or dismantling cast components:

....

....

....

When assembling vehicle components, different methods of sealing are used to ensure the best results in a given situation. This can vary from such as high pressures in the engine cylinder to high temperatures in the exhaust, or hot oil at high pressure in the engine.

Examine a complete engine gasket set and list the materials used for the different gaskets:

Gasket	Material
Cylinder head	

CORROSION

Corrosion, especially of bodywork, can seriously reduce vehicle life. What precautions should be taken to minimise corrosion due to service repairs?

....

....

....

Show the effect of corrosion after assembly when using components of dissimilar metals and indicate how corrosion can be avoided.

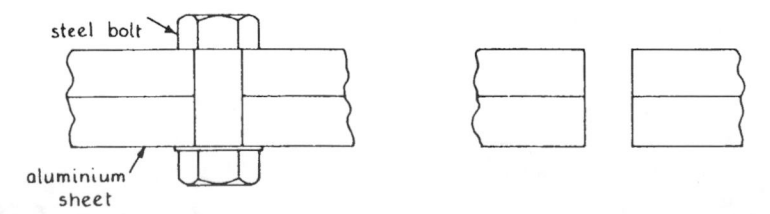

TYPES OF CORROSION

Corrosion is the natural tendency of metals to return to their oxidised states. Carbon steels are particularly vulnerable to corrosive attack.

State TWO common forms of corrosion:

1. ....

2. ....

Oxidation

Most metals are produced from the metallic oxide ores which are their natural state. Oxidation or chemical corrosion is a natural process which allows oxygen to combine with the surface of the metal and in the case of ferrous metals, slowly to form flakes of rust.

What are the main causes of chemical corrosion?

....

....

The effect chemical corrosion has on an unprotected vehicle chassis or body may be summarised as follows:

Oxygen attacks the surface and creates red iron oxide (rust). This makes the metal very porous and easily penetrated by oxygen, so further oxides form. The metal loses its strength and expands to a brittle layer which will chip away leaving the parent metal so thin that it can fracture.

The rusting will continue at a rate determined by the amount of moisture and chemical contaminants present. This can be considerable such as when damp dirt is lodged into crevices under wings.

When is the oxidation process considered to be an advantage?

....

....

....

....

....

....

Electro-chemical Effect

When two different metals are in contact in damp conditions the moisture will act as an electrolyte, similar to a battery cell, and the metal which has a negative electrical potential relative to the other metal will be corroded away. Any unprotected pair of different metals will corrode in this manner.

The table below shows the electrode potential of different metals:

Metal	Electrode Potential (volts)
Aluminium	−1.68
Zinc	−0.76
Chromium	−0.56
Iron/steel	−0.44
Cadmium	0.40
Nickel	−0.22
Tin	−0.14
Lead	−0.13
Copper	+0.40

Which material would corrode first if:

Steel were coated with zinc (galvanising)?

....

Aluminium and steel were bolted together?

....

Steel was plated with tin?

....

Sketches show what occurs when moisture penetrates steel that has been coated with:

(i) *Zinc*   (ii) *Tin*

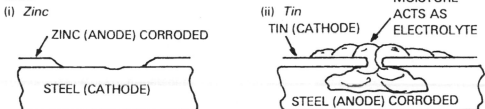

State FIVE effects of corrosion on a modern vehicle:

....

....

....

....

....

SURFACE PROTECTION

Surfaces are given various types of protective coatings. Sometimes these coatings are also decorative, for example, body paintwork.

What are the basic reasons for providing 'surface' protection?

....

....

....

One method of protecting materials is by coating the material with a thin layer of non-corrosive material. Some examples of this are shown below. State any motor-vehicle component protected by the processes named.

Protection process	Motor-vehicle component
Galvanising	
Tin plating	
Anodising	
Cadmium plating	
Chromium plating	

Spare parts which may be stored for a long time are protected with various substances. Examine some such components.

State how or what material is used to protect their surfaces.

Component	Type of protection

The vehicle body may be protected by various painting processes. With each process, different types of paint materials are applied to the surface.

What is the object of:

(a) The first surface primer-coat? ....

....

....

....

(b) Various layers of undercoat? ....

....

....

....

....

(c) Final high-gloss coat? ....

....

The underside of a vehicle is usually given extra protection.
Name and very briefly describe the three types of protection that may be used:

1. ....

2. ....

3. ....

List the main causes of corrosion on a vehicle:

1. ....

2. ....

3. ....

4. ....

5. ....

Chapter 10

Joining

JOINING OF MATERIALS AND COMPONENTS

The proper joining of parts or materials is essential in the construction of a modern motor vehicle. The parts may be fabrications such as vehicle bodies or assembled units of engines and gearboxes, etc. The choice of joining will be by the best and most economical method possible. TEN methods are listed below.
Give motor-vehicle examples of where such joining occurs.

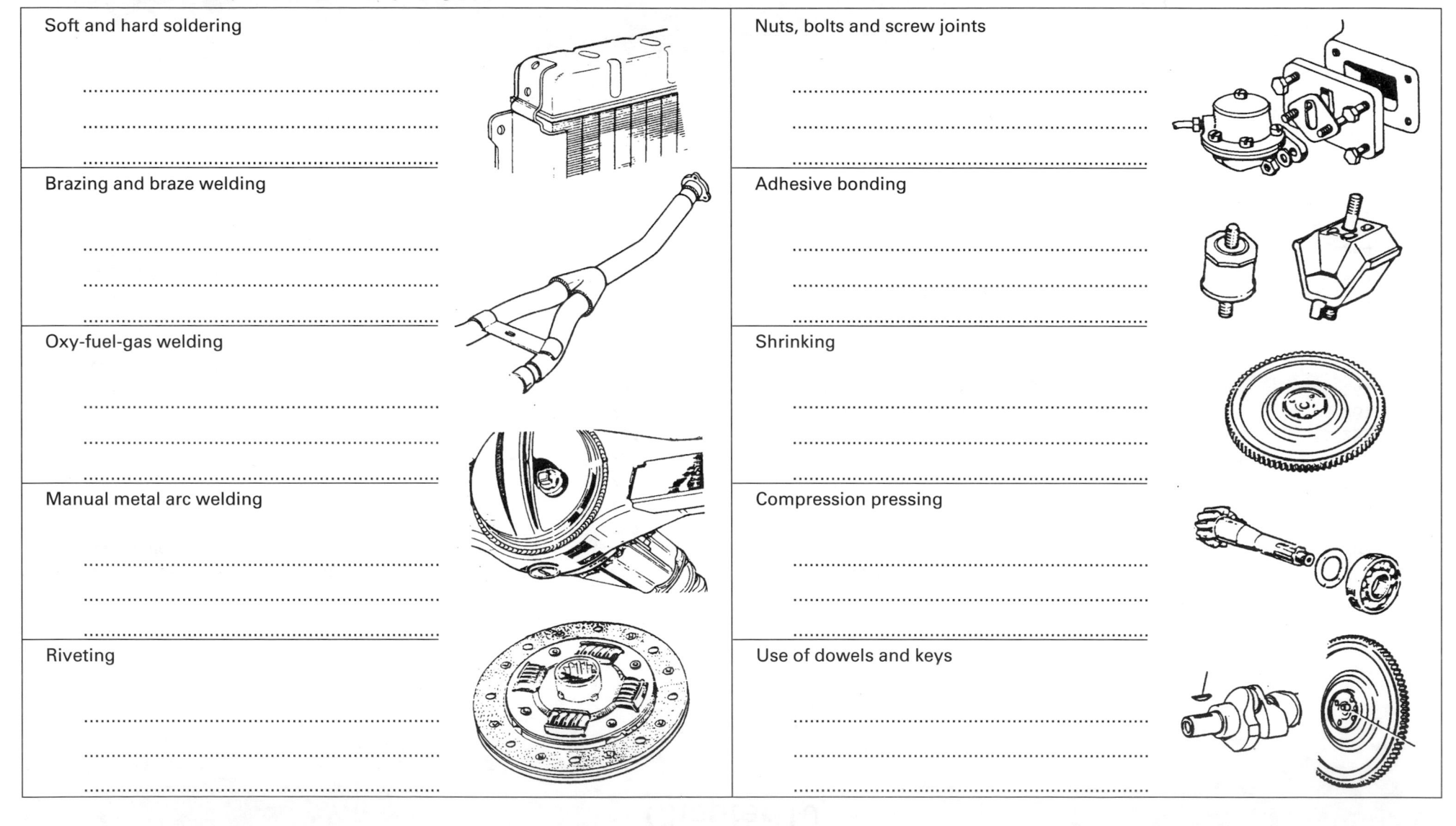

Soft and hard soldering

....

....

....

Brazing and braze welding

....

....

....

Oxy-fuel-gas welding

....

....

....

Manual metal arc welding

....

....

....

Riveting

....

....

....

Nuts, bolts and screw joints

....

....

Adhesive bonding

....

....

Shrinking

....

....

Compression pressing

....

....

Use of dowels and keys

....

....

SOFT SOLDERING

Soft soldering is a low-temperature metal-joining process used for joints which are relatively lightly loaded and not subjected to severe heating. It is also used for securing electrical cable connections and some pipe unions.

List FOUR metals that can be joined by soft soldering:

....

Soft soldering involves:

1. Preparing the surfaces to be joined.
2. Applying flux.
3. Applying sufficient heat for 'tinning' and running the molten solder into the joint.

How are the joint surfaces prepared?

1. ....

2. ....

3. ....

Why is this preparation necessary?

....

....

....

....

State the purpose of 'tinning' a soldering iron:

....

....

....

The heat required to warm the irons may be supplied in four ways:

1. ....   2. ....

3. ....   4. ....

Complete the drawing below to show an electrical soldering iron soldering the connection between a stator lead and diode wire in the alternator's rectifier pack.

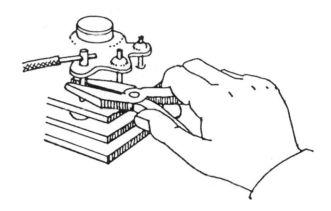

Complete the drawing below to show a soldering iron being held to the work at approximately the correct angle to create capillary action to sweat the joint.

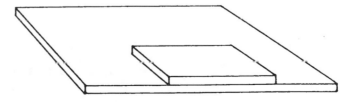

251

Solder

Soft solders are *lead–tin* alloys which may contain a small percentage of *antimony*. The percentages of lead and tin in a solder are varied to suit the application. As the percentage of lead in the solder is increased, the melting temperature range between initial melting and fully molten becomes greater, that is, the solder remains 'plastic' or 'pasty' over a wide temperature range. At what temperature does ordinary solder *begin* to melt?

....

Complete the table below:

% Lead	% Tin	Temperature range while solder is in a plastic state	Typical use
40	60		
50	50		
70	30		

50/50 solder is commonly called ....

40/60 lead–tin solder is commonly called ....

70/30 lead–tin solder is commonly called ....

State the advantage of 'tinmans' solder

....

....

....

What benefit is to be gained by using a solder which has a prolonged 'pasty' stage when solidifying?

....

....

....

Fluxes

State the purpose of a soldering flux:

....

....

....

The type of flux must be chosen to suit the application. Two types of flux are in general use.

State materials for which the following fluxes are best suited:

Corrosive	
Hydrochloric acid	
Zinc chloride (killed spirit)	

Non-corrosive	
Tallow	
Resin	

Corrosive fluxes are .... fluxes which prevent oxidation taking place, thereby protecting the joint surfaces during soldering. These fluxes also help to clean the joint surfaces.

Non-corrosive fluxes are .... fluxes which protect the 'cleaned' surface during the soldering process.

What precautions should be taken when using corrosive fluxes?

....

....

....

....

Soldering a Joint

The joint shown above is a .... joint.

Outline the procedure for soldering this joint.

(a) ....

(b) ....

(c) ....

....

This soldering process is known as ....

It may be necessary to apply a little solder at 'B'. If this is done, the solder will flow into the joint formed between the faces; this is known as the

.... action of the solder.

Heat loss can be a problem when soldering. How can this be minimised?

....

....

Another problem associated with soldering is the damaging effect of heat flow to certain parts of a component being soldered. It is possible that heat flow to special regions can be limited during the soldering process by strategically positioning blocks of metal on the workpiece or holding wire with pliers to act as heat sinks and absorb the flow of harmful heat.

List THREE common soldering faults:

....

....

....

HARD SOLDERING

With any soldering process, the metal used for joining melts at a lower temperature than does the parent metal. In this connection brazing is a hard soldering process. Another hard soldering process is 'silver soldering'.

What are the essential similarities/differences of hard and soft soldering?

Similarities ....

....

....

Differences ....

....

....

....

What are the basic ingredients of 'silver solder' and how is the silver soldering process carried out?

Ingredients ....

Process ....

....

....

Give three reasons for joining metals by silver soldering:

1. ....

....

2. ....

....

3. ....

BRAZING AND BRAZE WELDING

In each case the base metals being joined are brought to a red heat, but not actually melted. The filler metal used to effect the joint (normally a brass alloy) has a melting point above 450°C (to distinguish it from soldering), but below that of the base metals.

Brazing

The filler metal distributes itself into the joint by capillary action. Joint designs are similar to those used for soldering.

State the reason why brazing would be used in preference to fusion welding:

....

....

....

....

....

What is the most common brazing fault?

....

....

....

Braze Welding

Bonding occurs in a similar manner to brazing, but the filler metal is NOT distributed by capillary action. Instead it is added to the joint as welding rod would be, or is deposited by arc welding. Joint designs are similar to those used for oxyacetylene welding.

Give one example of braze welding in motor-vehicle repair, and state the reason for its use:

....

....

....

....

OXYACETYLENE WELDING

This form of welding is still widely used in general repair work, but more specialised applications demand more modern types of welding equipment. During welding the metals being formed are actually melted so that they flow together.

Acetylene is a hydrocarbon fuel; it reacts with oxygen to liberate heat at high temperatures.

Oxygen: A stable gas, contained at high pressures of about 172 bar to 230 bar, in black-painted cylinders.

Acetylene: An unstable gas, contained at medium pressures of about 15.5 bar to 17.7 bar, in cylinders packed with kapok, or kapok and charcoal. Acetylene cylinders are painted maroon.

Complete the drawing below showing a typical oxyacetylene welding set. Name the parts and state the colour of the cylinders.

State THREE faults common in this type of welding:

....

....

....

MANUAL METAL ARC WELDING

An electric arc is a sustained spark caused by current jumping across the gap between two terminals in an electrical circuit. In metallic arc welding, the arc is formed between the metal to be welded and an electrode. The arc creates intense heat in a small area. This melts the metal and the tip of the electrode; molten metal from the electrode then transfers across the arc to act as filler.

Equipment

The power source that supplies the current for metal arc welding may be either a.c. (alternating current) or d.c. (direct current). With a.c. the energy is supplied from an a.c. welding transformer set. The basic equipment for d.c. comprises a generator which may be driven by an electric motor or an internal combustion engine.

Name the parts indicated on the sketch:

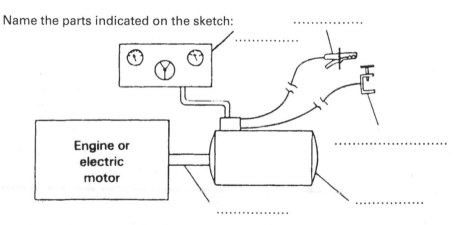

A direct current (d.c.) welding set

What are the particular advantages and disadvantages of manual metal arc welding as used in the vehicle-repair trade?

....

....

....

....

RESISTANCE WELDING

Resistance welding is a process in which the welding heat is generated by the resistance offered to the passage of an electric current through the parts being welded.

The form of resistance welding widely used on motor vehicles is known as 'spot welding'. This is a process where the two metal sheets being joined are clamped between two large copper electrodes which pass the welding current through the plates.

Name the main parts of the spot welder shown below:

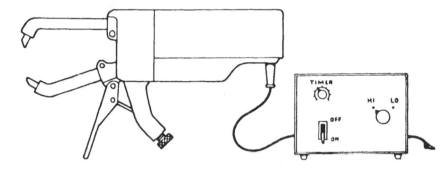

State the advantages of this method of joining metals:

....

....

Three forms of resistance welding are spot, seam and projection (protrusion). Seam uses wheels and produces a watertight joint.
Protrusion presses on projections which collapse during welding (securing clips are spot welded by this process).

Investigation

Examine a modern vehicle and state TWO places where each resistance weld is used:

Spot	Seam	Projection
1.	1.	1. ....
2.	2.	2. ....

SHIELDED ARC PROCESS

With the metallic arc process already described, oxidisation is prevented by slag-forming fluxes in the electrode coating. An alternative method of preventing oxidisation is to shield the arc and molten metal from the atmosphere by shrouding the area with an inert gas, i.e. a gas that will not combine with heated metals. The gas argon is widely used for this process.

Argon Arc – Tungsten Inert Gas Shielded (TIG)

The heat for welding is supplied by an arc struck between a 'non-consumable' tungsten electrode and the joint to be welded. The argon gas shield is supplied through a nozzle concentric with the electrode tip. Most joints require the use of a filler rod.

The diagram below shows the layout of the tungsten argon arc welding plant; label the diagram.

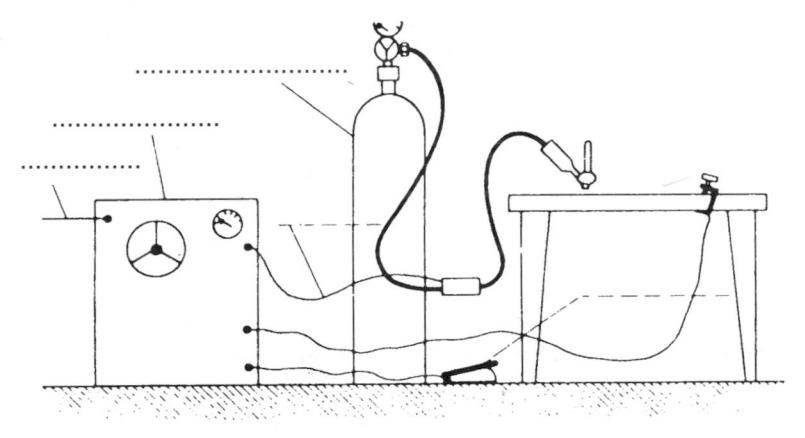

State three advantages of the TIG system:

1. ....

....

2. ....

....

3. ....

CO$_2$ Process – Metal Arc Gas Shielded (MAG)

In this process a 'consumable' electrode in the form of a wire is fed automatically from a reel into the arc, the weld area being shrouded with CO$_2$ gas.

The equipment for CO$_2$ welding comprises the following essential parts: a d.c. power unit, a CO$_2$ gas supply, a gun or torch to which the filler wire, current and gas are fed, and a control unit for the reel of filler wire.

This diagram shows the layout of equipment for CO$_2$ welding; label the diagram.

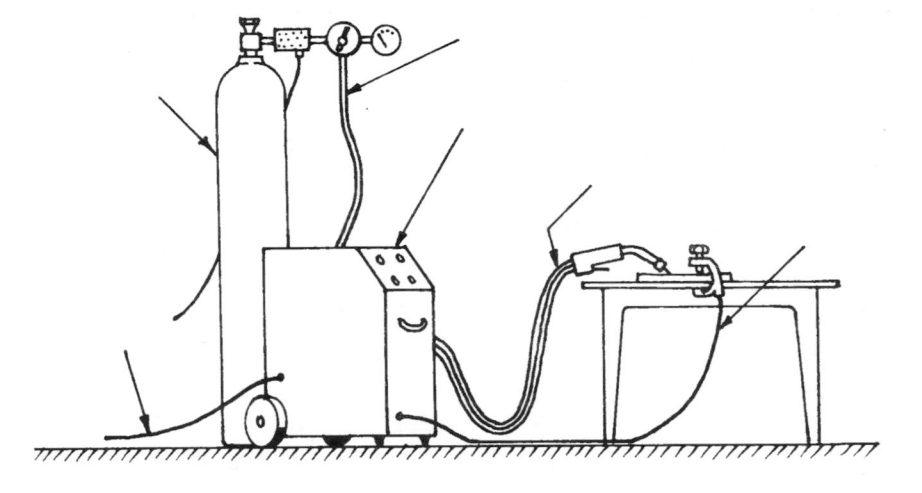

This type of welding is a popular form of welding in Auto Body repair.

State the advantages of the MAG system when compared with other forms of welding:

....

....

....

RIVETING

Riveting is one form of mechanical fastening which makes a permanent joint between two or more materials such as chassis frames, body fabrications, or metal to fabric.

State FOUR materials from which rivets are commonly made:

....

A solid rivet is shown below. Make sketches above each title to show the types of rivets named.

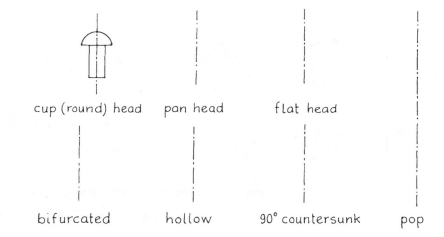

cup (round) head   pan head   flat head

bifurcated   hollow   90° countersunk   pop

Pop Rivets

Pop rivets are hollow rivets which are supplied already threaded on to a mandrel. They are closed by gripping the mandrel in a riveting tool and pulling it so that it forms a head on the remote end of the rivet and then breaks off.

What are the main advantages of pop rivets?

....

....

....

....

The sketch at A below shows the pop rivet and the riveting tool in position before the rivet is closed. Complete the sketch at B to show the shape of the rivet after closing.

A   B

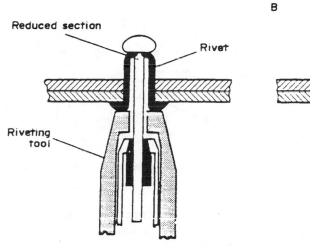

Reduced section

Rivet

Riveting tool

Some pop rivets have a specially shaped head which remains trapped in the formed head. Such a rivet is shown below; why are these used?

....

What type of rivets are used for attaching brake and clutch linings to the relative faces, and what is the reason for their use?

....

....

....

....

....

JOINING BY NUT AND BOLT OR SCREW

The nut and bolt is the most common method of joining two detachable components together. Name the parts of the nut and bolt shown.

Bolts normally have a hexagon or allan socket type of head while the screw thread is usually of a Vee type. A screw has a screwdriver type of head fitment and a thread similar to a bolt, or a faster thread similar to a woodscrew or self-tapping screw.

What types of bolt heads other than hexagon are used on modern vehicles?

1. .... 2. ....

Strength and grade identification

State what the symbols on the bolts etc. indicate:

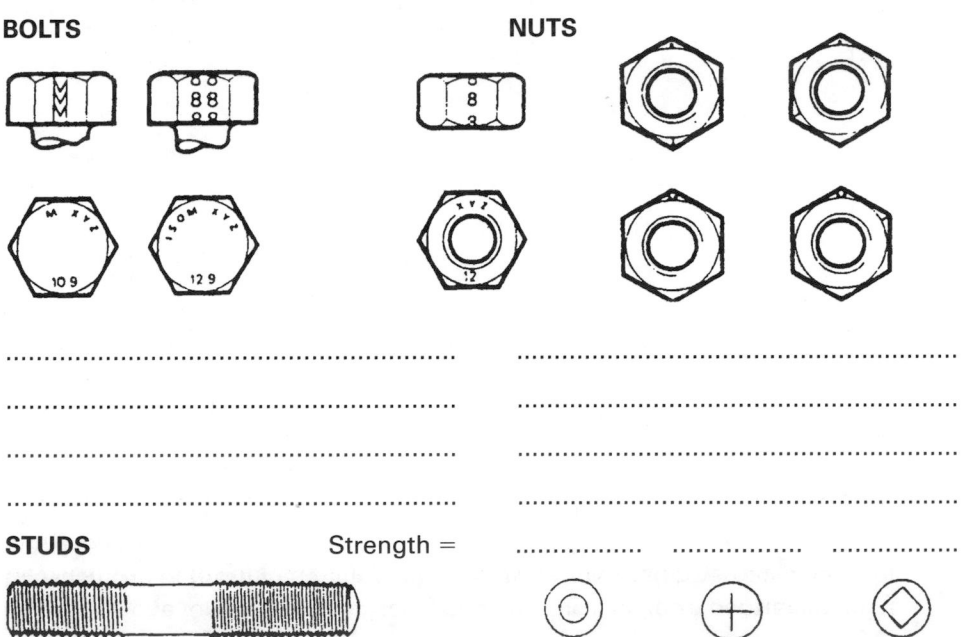

BOLTS   **NUTS**

....
....
....
....

STUDS   Strength = ...

Types of screw heads

Name the types of screw heads shown below:

....

Types of head to suit screwdriver

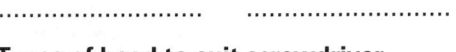

....

State the advantage of crosshead screws:

....
....
....

Clearance in bolt holes

All bolt and stud holes must be given a correct amount of clearance to ensure that a joint on a component can be easily separated. Why may it be difficult, after some time, to separate a joint that has been securely tightened?

....

The drawings below show this clearance.

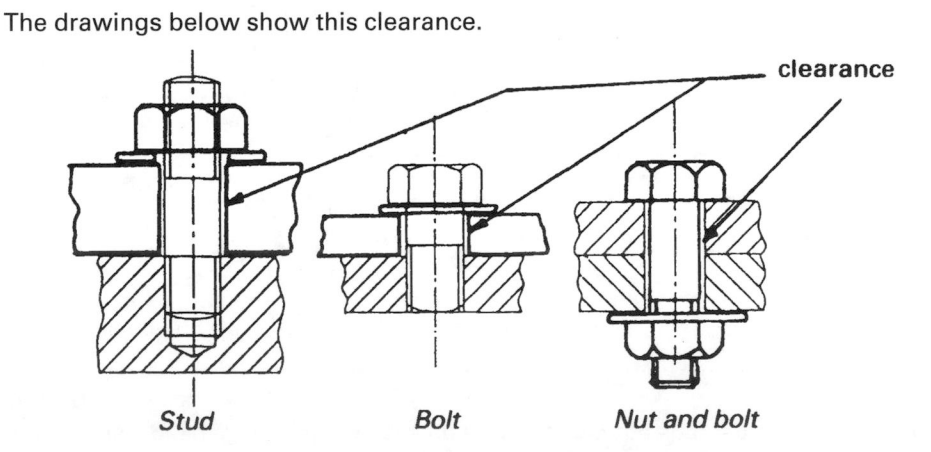

Stud   *Bolt*   *Nut and bolt*

LOCKING DEVICES

Most of the nuts and bolts used on motor vehicles are fitted with locking devices. This is to counteract the loosening effect caused by strain and vibration.

Name the types of locking device shown below and give an example of where each may be used on a motor vehicle:

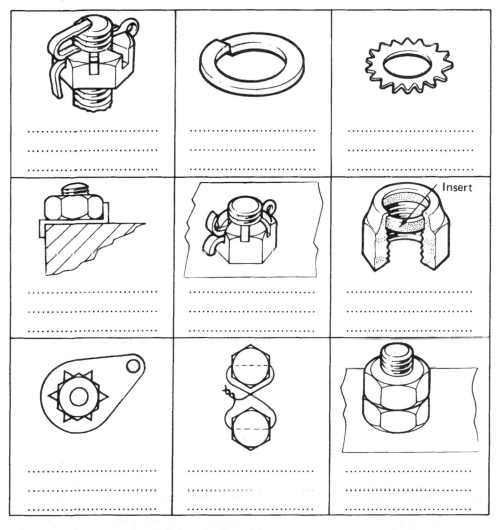

Which of the locking devices shown opposite are

Positive locking devices? ....

....

Frictional locking devices? ....

....

Describe an alternative method of locking nuts and bolts that is very often used in engine assembly:

....

....

Both 'left-' and 'right'-hand screw threads are used on motor vehicles. However in most cases it is the right-hand thread which is used.

State which of these represents left- and right-hand threads:

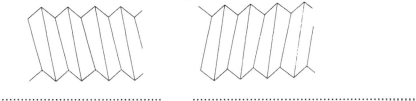

....

State two places on a motor vehicle where left-hand threads are used and give the reasons for using them:

1. ....

....

2. ....

....

When tightening nuts and bolts, care must be taken to ensure that they are not left slack or overtightened. What tool should be used to ensure a correct tightness is applied?

....

What problems are associated with the overtightening of nuts or bolts?

....

....

USE OF ADHESIVES

A considerable range of adhesives is used in the automotive industry today. Such adhesives have a large variety of applications and in many instances are replacing joints previously riveted or welded.

Requirements

The requirements of an adhesive are that it will 'wet' and adhere to the surfaces being bonded and that the film in between the surfaces will have a high cohesive strength.

Adhesive Types

Thermo-hardening adhesives

Thermo-hardening adhesives will form a strong heat-resistant joint between a wide variety of materials. The adhesive is made up from two parts – a resin and a hardener (catalyst curing agent) – which when mixed transform into a hard solid. These adhesives, which contain epoxy resin, require heat to effect a 'cure'.

State the meaning of the following terms as applied to adhesives:

Structural adhesives ....

....

Thermo-plastic ....

....

Impact type ....

....

Hot set ....

....

Cold set ....

....

There are many advantages of adhesives, three of which are listed below. State three more advantages and three disadvantages.

Advantages

1. *They give stress distribution over the whole bond area.* ....

2. *They can be used to joint a wide variety of materials.* ....

3. *Ideal for joining fragile materials.* ....

4. ....

5. ....

6. ....

Disadvantages

1. ....

    ....

2. ....

    ....

3. ....

    ....

State the meaning of the term 'wetting the surface' in relation to adhesives:

....

....

....

To what part of the joining operation does the word 'cure' relate?

....

....

It is extremely important when using epoxy resins that the correct procedure is adopted. What factors must be considered before and when making a joint so as to avoid faults?

(a) ....

(b) ....

(c) ....

(d) ....

(e) ....

Give THREE examples of the use of epoxy resins in the construction or repair of automobiles:

1. ....

2. ....

3. ....

Joining Process – Thermo-hardening Adhesive

Outline the sequence of operations for making a joint between mild steel and aluminium, using any suitable adhesive:

Type of joint ....

Adhesive trade name ....

....

....

....

....

....

....

Joining Process – Impact Adhesive

Impact adhesives contain a solvent which, when exposed to air, evaporates.

The surfaces being joined, having been suitably prepared, are coated with the adhesive and after a period of time (stipulated by the manufacturer) are brought into contact and pressure applied. These adhesives, which are often supplied in tape form, are generally less rigid and weaker than thermo-hardening types.

Give six examples of the use of impact adhesives in the construction or repair of automobiles:

1. ....   2. ....

3. ....   4. ....

5. ....   6. ....

Name four popular impact adhesives which are used in the motor trade and state a typical use for each:

1. ....

2. ....

3. ....

4. ....

The cyanoacrylate adhesives are impact adhesives. These are very quick acting

adhesives and are known as ....

Safety precautions when using adhesives

State the safety precautions to be observed when using adhesives:

....

....

....

....

FITTING PARTS BY SHRINKING

The component to be fitted is first either heated and expands or is cooled and contracts before fitting. In all cases the components must be machined very accurately to the correct sizes.

Hot Shrink

The part is usually an external gear or collar and has a smaller internal diameter than the diameter on to which it is to be fitted. The most common example of a hot shrunk component on a vehicle is the ring-gear fitted to the engine flywheel. Complete the table to show the ring-gear internal diameter.

Heating process	Ring-gear internal diameter	Flywheel external diameter
Before heating		320 mm
Ring-gear heated to 300°C		320 mm
When fitted and cooled		320 mm

Describe how a worn ring-gear can be removed from the flywheel:

....
....
....

What is the correct heating procedure to fit a new ring-gear?

....
....
....
....

What technique should be applied when fitting the ring-gear to the flywheel?

....
....
....

Cold Shrink

The component is usually an internally fitted part which is larger than the hole into which it is to be located. An engine valve seat insert is a common motor-vehicle example.

Sketch shows a valve seat insert about to be fitted into a cylinder head.

VALVE SEAT INSERT

The shrinkage process is more commonly carried out by the manufacturers who have the equipment capable of cooling to temperatures of −190°C.
How can a component be cooled in a motor-vehicle workshop?

....
....
....
....
....

What are the advantages of hot or cold shrinking a component?

....
....
....
....
....

What faults may occur during the hot shrinking, cold shrinking or the fitting process?

....
....
....

FITTING PARTS BY COMPRESSION PRESSING

A compression joint is a joint requiring an interference fit which requires assembly by some form of pressure.

The fit is not as tight as a shrunk fit but is sufficiently tight. Indicate how the bush may be simply pressed into the casing.

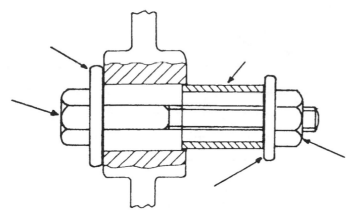

Explain how the tool is extracting the bush from the assembly.

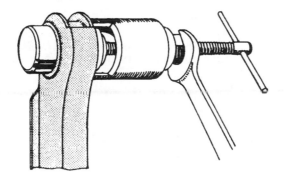

....

....

....

....

All compression joints have a centring edge.
What does this mean and what is the edge's function?

....

....

....

....

Before applying pressure to the joint, what preliminary procedures should be made?

....

....

....

....

What are the most common faults that occur when fitting components by compression pressing?

....

....

....

....

The five drawings below show a wheel hub assembly. State in each case what compression pressing procedure is occurring.

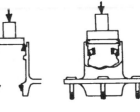

....

....

....

263

USE OF PINS AND KEYS

All types of pins and keys are items which locate parts before securely fixing them, preventing excessive movement or completely locking two parts together.

Pins

Identify the types of pin shown below:

....

A dowel is a short parallel shank pin which fits into a precisely reamed hole.
What are the functions of the dowels shown below?

....

....

....

....

....

....

....

....

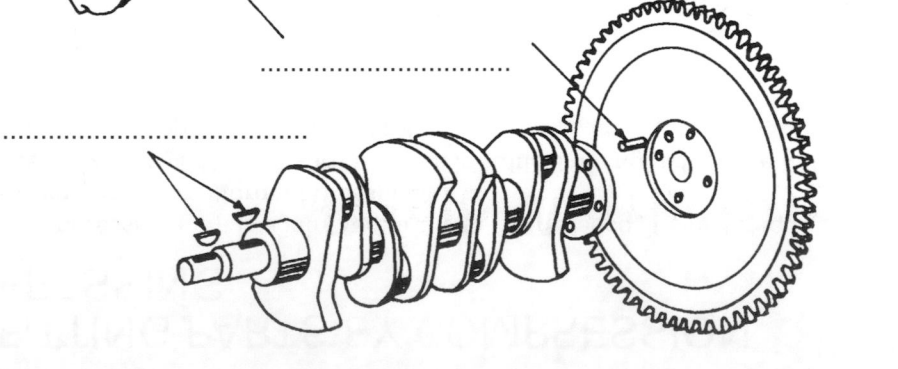

....

....

Taper and parallel pins

Taper pins fit into holes of similar taper. Parallel pins, including split pins, allow easy dismantling.
What are the advantages of using taper pins?

....

....

What is the main problem encountered when fitting pins?

....

....

The pin shown below is called a ....
It has a nut at one end to secure the
pin in position.
Give TWO examples of where on a
motor vehicle such a pin may be
used:

1. .... 2. ....

Keys

Keys prevent pulleys and gears on shafts from turning.
Identify the keys shown below:

....

The first key is the one most commonly found on motor vehicles.
Describe how the two keys positioned to fit in the crankshaft opposite should be correctly assembled together with the timing chain pulley and crankshaft pulley:

....

....

....

....

....

....

Chapter 11

Basic Workshop and Vehicle Calculations

This chapter is included in order to help students become familiar with the simple, everyday type of calculations that they are likely to use in connection with motor-vehicle work and leisure interests. It also aids the Key Skill Areas such as Application of Number and Problem Solving.

AVERAGES

The average (or mean value) of a set of numbers is equal to their sum divided by the number of them.

Find the average of 3, 5, 9 and 11

Average $= \dfrac{3 + 5 + 9 + 11}{4} = \dfrac{28}{4} = 7$

During a 4 hour journey a man did 40 km in the first hour, 48 in the second, 44 in the third and 51 in the last hour. What was his average speed for the journey.

Average speed $= \dfrac{40 + 48 + 44 + 51}{4}$

$= \dfrac{183}{4} = 45\frac{3}{4}$

$= 45\frac{3}{4}$ km/h

Problems

1. A vehicle's fuel consumptions during three different journeys of identical mileages were calculated as 11.5 km/litre, 12.5 km/litre and 11.25 km/litre. The average fuel consumption figure for the vehicle during the three journeys of equal mileage is therefore

   (a) 10 km/litres   (b) 11.25 km/litres

   (c) 11.75 km/litres   (d) 15 km/litres

   Answer (   )

2. During a check on four spark plugs, the gaps were found to be 0.6 mm, 0.55 mm, 0.7 mm and 0.75 mm. The average gap was

   (a) 0.7 mm   (b) 0.8 mm

   (c) 0.6 mm   (d) 0.65 mm

   Answer (   )

3. If a car averages 62 km/h for $4\frac{1}{2}$ hours, the total distance covered will be

   (a) 93 km   (b) 124 km

   (c) 279 km   (d) 379 km

   Answer (   )

4. In five consecutive months the number of services done by a garage was 86, 90, 79, 74 and 81. The monthly average was

   (a) 88   (b) 44

   (c) 80   (d) 82

   Answer (   )

5. Four different oil measures have capacities of 4 litres, 2 litres, 1 litre and $\frac{1}{2}$ litre. The average capacity is

   (a) $1\frac{1}{2}$ litres   (b) $2\frac{1}{2}$ litres

   (c) 3 litres   (d) $1\frac{7}{8}$ litres

   Answer (   )

6. After a tyre inspection on a vehicle the tread depths were found to be 3.5 mm, 2.5 mm, 0.5 mm and 2 mm. What was the average tread depth?

   ....

   ....

   ....

   ....

7. If the average interval between three oil changes on a vehicle is 9000 km, and the intervals between the first two are 6000 km and 10 000 km, what is the interval between the second and third oil change?

   ....

   ....

   ....

   ....

8. If a roll of copper tubing 16 m in length costs £86.40 what is the average cost per metre?

   ....

   ....

   ....

   ....

9. Six apprentices receive a wage of £40 each. If four other apprentices receive wages of £48 each, what is the average wage of all the apprentices?

   ....

   ....

   ....

   ....

10. If the times taken to fit water pumps to three cars of the same type are 2.5 hours, 2.8 hours and 2.4 hours, what is the average time for the job?

    ....

    ....

    ....

    ....

11. If the total number of crankshafts sold by the stores in one year is 276, what is the average monthly sales figure?

    ....

    ....

    ....

    ....

PERCENTAGES

Per cent or percentage (usually denoted by %) refers to a part or fraction of some quantity expressed in hundredths.

To obtain the percentage, the fraction or the decimal is multiplied by 100.

Express the following as percentages:

(a) $\frac{3}{5} \times \frac{100}{1} = 60\%$ (b) $\frac{9}{75}$ (c) $\frac{23}{32}$

(d) 0.4   (e) 0.35   (f) 0.01

When commencing a workshop exercise, 15 students share equally a bar of mild steel 2 m in length. What percentage of the bar will each student receive?

Each will receive $\frac{1}{15}$ of the bar.

∴ The percentage will be $\frac{1}{15} \times \frac{100}{1} = 6\frac{2}{3}\%$

If the repair charge for a job is £120 and 27% of this is the cost of parts, how much did the parts cost?

$$\text{Cost of parts} = \frac{27}{100} \text{ of £120}$$

$$\therefore \text{Cost of parts} = \frac{27 \times 120}{100} = \frac{162}{5} = 32\frac{2}{5}$$

$$\text{Cost of parts} = £32.40$$

1. A vehicle's cooling system capacity is 8 litres and a mixture of 25% antifreeze is required. How much antifreeze must the owner purchase?

2. If 2 litres of antifreeze are mixed with water in a 4 : 1 water/antifreeze ratio, how many litres of mixture are available to fill a cooling system and what percentage antifreeze does this mix represent?

3. The rate for repair charges at a garage is £18.15 per hour. If 20% of this is profit, 60% is overheads and the rest is the mechanic's pay, at what hourly rate is the mechanic paid?

4. A stick of solder has a mass of 3 kg. If it is composed of 30% tin and 70% lead, what is the mass of lead?

5. An engine cylinder is to be rebored to give a 2% increase in diameter. If the original diameter is 60 mm, determine the diameter after reboring.

6. If the retail price for a brake master-cylinder is £30 and a mechanic is allowed $12\frac{1}{2}\%$ discount on this, how much would he pay for the master cylinder?

7. If a car reduces speed by 8% and is then travelling at 70 km/h, what was its former speed?

8. The cooling system capacity of a commercial vehicle is 15 litres and a 30% antifreeze mixture is to be used. How much antifreeze is required?

RATIO AND PROPORTION

A 'ratio is a comparison of the size of two quantities. For example, if the stroke of an engine is 80 mm and the bore of the engine is 50 mm, the stroke-to-bore ratio would be:

Ratio $= \frac{80}{50} = \frac{8}{5}$ or $1\frac{3}{5} : 1$

Divide a line 48 mm long into two parts in the ratio of 3 : 1 and determine their lengths.

1 part $= \frac{48}{4} = 12$

3 parts $= 12 \times 3 = 36$

Therefore the two parts would be 36 and 12 mm.

Problems

1. The ratio of petrol and oil for a certain two-stroke engine is 20 : 1. Therefore the oil added per litre of petrol would be

   (a) 0.1 litres   (b) 0.2 litres

   (c) 0.05 litres   (d) 0.02 litres

   Ans. (   )

....

....

....

....

....

....

....

'Proportion' is the relationship between two or more quantities, i.e. if one quantity was changed, the others would correspondingly increase or decrease to keep the relationship the same. For example, if 2 litres of oil cost 15p, how much would 6 litres cost?

2 litres cost 15p

1 litre costs $\frac{15p}{2}$

Therefore 6 litres would cost

$\frac{15}{2} \times \frac{6}{1} = 45p$ *Ans.*

2. 14 litres of fuel are pumped into a tank in 25 seconds. How much fuel would be pumped into it in 45 seconds?

....

....

....

....

....

....

....

3. Calculate the rear axle ratio of a vehicle, if the crown wheel has 30 teeth and the meshing pinion 7 teeth.

....

....

....

....

4. A mixture of antifreeze and water is in the ratio of 4 parts water to 1 part antifreeze. How much antifreeze should be added to a cooling system with a capacity of $6\frac{1}{2}$ litre?

....

....

....

....

5. A commercial vehicle chassis is 11.5 m in length. If the chassis is lengthened in the ratio of 5 : 4, what is the new length of the chassis?

....

....

....

....

6. A hydraulic ramp lifts a vehicle to a height of 2 m in 30 seconds. At what height would the vehicle be after 13 seconds?

....

....

....

....

7. A mechanic, using a spanner 175 mm long, exerts a torque on a bolt of 14 N m. What torque would be produced by applying the same force on a spanner 20 mm longer?

....

....

....

....

8. Five similar cars, doing similar work, use 150 litres of fuel in three days. How long would the same quantity of fuel last two of the cars?

....

....

....

....

AREAS OF REGULAR FIGURES

The figure shown below is a square. That is, length and breadth are equal.

The shading represents the area of the square.

To calculate the area, we square the length of one side.

area $= a \times a = a^2$

Rectangle

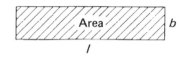

The area of a rectangle as shown above is found by multiplying the length and breadth.

area $= l \times b$

State the SI derived units used to express area ....

....

Perimeter

How is the perimeter of a square rectangle calculated?

....

....

Examples

1. Calculate the area of a square whose sides are 3 m in length

Area $= 3 \times 3 = 9$ m^2

2. Calculate the area of the rectangle shown below.

35 mm

150 mm

Area $= 150 \times 35 = 5250$ mm^2

3. A workshop is 12 m square. The floor area is therefore:

   (a) 12 m^2 (b) 24 m^2

   (c) 144 m^2 (d) 240 m^2

   Answer (   )

4. A workshop has 4 work benches each measuring 1 m $\times$ 2$\frac{1}{2}$ m. The floor area covered by the benches is therefore (a) 4 m^2, (b) 5 m^2, (c) 2$\frac{1}{2}$ m^2, (d) 10 m^2

   Answer (   )

5. A motor-lorry platform measures 10 m $\times$ 2 m. How many boxes whose bases measure 0.5 m $\times$ 0.5 m would it take to cover the platform?

....

....

....

....

....

6. The figure shown below represents the workshop floor and pit area. Calculate

   (a) the area of the pit, and

   (b) the area of the remaining floor space.

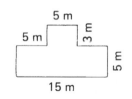

....

....

....

....

....

....

....

....

7. Determine the area and perimeter of the following shapes.

(a)

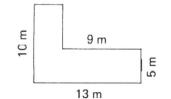

(b)

8. Complete the following table, in respect of rectangles.

Length	Breadth	Area
4.5 m	3 m	
150 mm	100 mm	
8 m		16 m^2
	3.5 m	21 m^2
25 mm	200 mm	

Triangles

All triangles have an area which is half that of a rectangle whose length and breadth is equal to the base and perpendicular height of the triangle. For example.

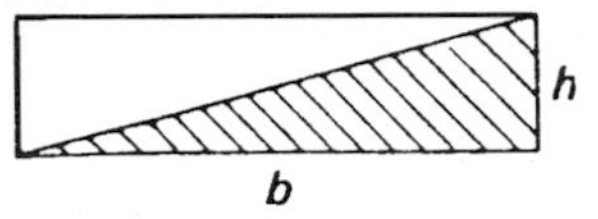

Therefore the area of a triangle would be

$$\text{Area} = \frac{\text{base} \times \text{perpendicular height}}{2}$$

This holds good for any triangle provided the height used is the perpendicular height.

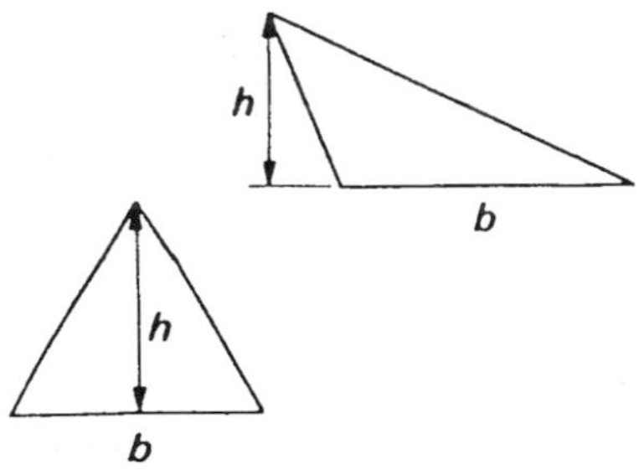

Circle

To calculate the area of a circle the following formula can be used:

$$\text{Area} = \pi r^2 \text{ or } \frac{\pi d^2}{4}$$

Where

$$\pi = \frac{22}{7} \text{ or } 3.142$$

r = radius and d = diameter

The ratio of $\frac{\text{circumference}}{\text{diameter}}$ will always give a value of 3.142.

The symbol for this ratio is the Greek letter π (pi).

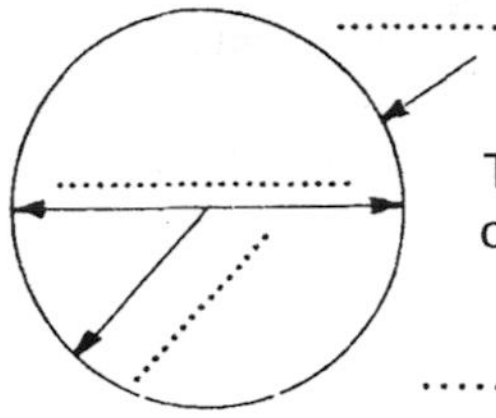

The 'perimeter' of a circle is the

....

State the formula to calculate the circumference of a circle.

Examples

1. Calculate the area of a circle which has a diameter of 40 mm.

$$\text{Area} = \pi r^2$$

$$\text{Area} = 3.142 \times 20 \times 20$$
$$= 1256.8 \text{ mm}^2$$
$$= 12.568 \text{ cm}^2$$

2. Calculate the area of the triangle shown below.

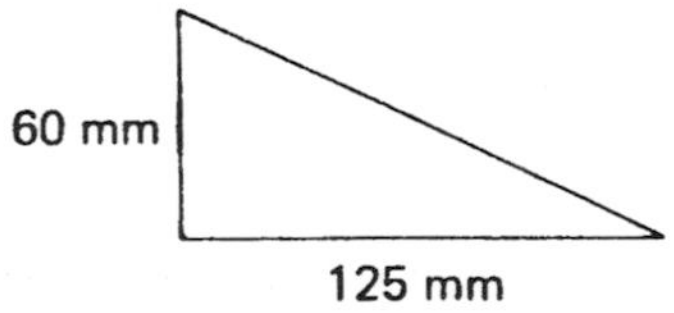

$$\text{Area} = \frac{b \times h}{2}$$

$$\text{Area} = \frac{125 \times 60}{2} = 3750 \text{ mm}^2$$
$$= 37.50 \text{ cm}^2$$

---

3. Calculate the area of a triangle which has a base of 20 cm and a perpendicular height of 45 cm.

....
....
....
....
....
....

4. A brake master-cylinder has a bore diameter of 25 mm. What is the area of the bore?

....
....
....
....
....
....

5. Determine the area of the shape

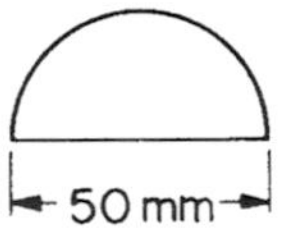

....
....
....
....

---

6. Determine the area and perimeter of the following shapes.

(a)

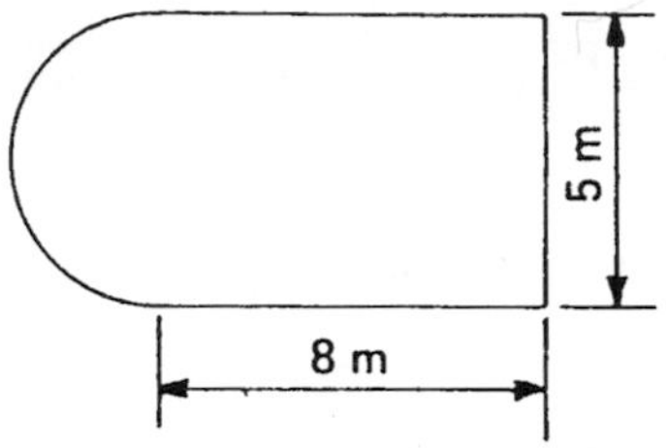

....
....
....
....
....

(b)

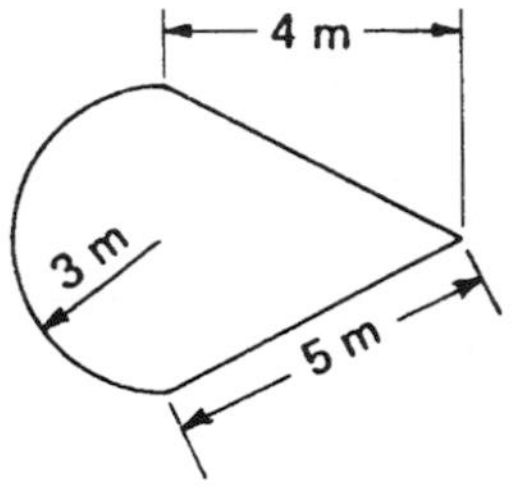

....
....

VOLUME OF REGULAR FIGURES

The volume of any regularly shaped object can be found by multiplying the area of the end by the length.
Examples of regularly shaped objects are:

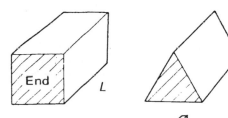
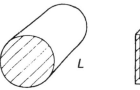

State three SI derived units commonly used to express volume.

....

....

....

If a figure is described as being regular shaped or of 'uniform' cross-section, what do these terms mean?

....

....

....

....

Examples

1. Determine the volume of the rectangular prism shown below.

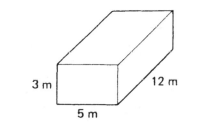

Volume = area of end × length
= 3 × 5 × 12
= 180 m³

2. If the cross-sectional area of a metal bar of uniform section is 0.0025 m² and its length is 4 m, the volume of metal would be

(a) 0.005 m³   (b) 0.01 m³

(c) 125 mm³   (d) 150 mm³

Answer (   )

3. If a cube has sides measuring 5 mm × 5 mm × 5 mm, its volume would be

(a) 25 mm³   (b) 50 mm³

(c) 125 mm³   (d) 150 mm³

Answer (   )

4. Calculate the volume of the rectangular prism shown below.

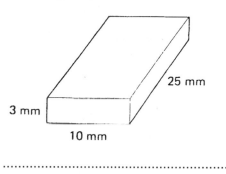

....

....

....

....

5. Calculate the volume of the section of angle iron shown below.

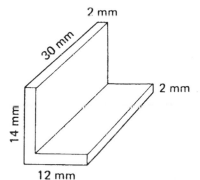

....

....

....

....

6. Determine the volume of the wedge shown below.

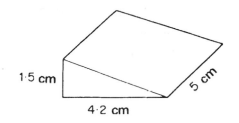

....

....

....

....

7. Calculate the volume of the compressor mounting block shown below.

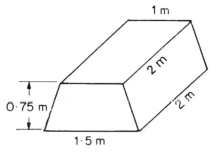

....

....

....

....

MASS

All substances consist of matter or molecules packed together to form a material, be it made of steel, wood, glass, plastic etc.

Mass is defined as ....

The SI unit of mass is the ....

FORCE

Force cannot be seen or touched but its effects can be observed, that is, if a force applied to a stationary object causes that object to move it would move in the ....

The effect that a force has on an object will depend upon:

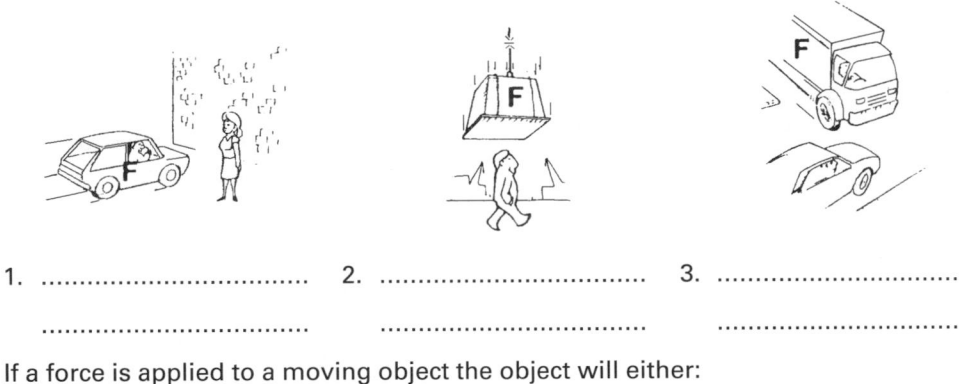

1. ....   2. ....   3. ....

   ....

If a force is applied to a moving object the object will either:

1. ....   2. ....   3. ....

ACCELERATION

Velocity is the distance travelled in a unit of time, for example, the number of metres travelled in one second (m/s).
When an object moves and steadily increases its velocity, for example, when a car drives away from rest, the object is said to have ....

Acceleration is ....

The SI unit of acceleration is ....

The relationship between force, mass and acceleration is found directly from the effects that a force produces.

CAPACITY

The capacity of a container or tank is the cubic content or volume that it will hold, When stating the capacity of a tank in connection with a quantity of liquid,

the units are .... Capacity is determined by calculating the volume and converting the cubic measurement into litres.

$$1 \text{ litre} = 1000 \text{ cm}^3 \qquad \text{or} \qquad 1 \text{ litre} = 0.001 \text{ m}^3$$

1. The volume of a tank is 1 m^3, its capacity in litres is therefore ....

   (a) 1   (b) 100   (c) 1000   (d) 10 000

2. Calculate the capacity of the fuel tank shown below (ignore capacity of filler neck and the rounded corners of the tank).

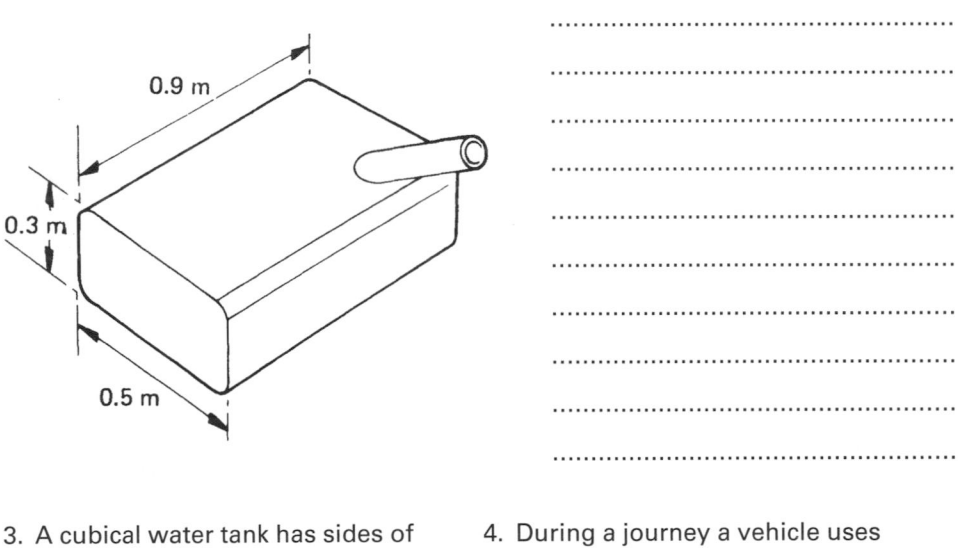

0.9 m
0.3 m
0.5 m

....
....
....
....
....
....
....
....
....
....
....
....

3. A cubical water tank has sides of 0.4 m, its storage capacity is therefore ....

   (a) 4 litres   (b) 12 litres

   (c) 64 litres   (d) 640 litres

4. During a journey a vehicle uses 30 litres of fuel from a tank. If the fuel tank has a volume of 0.045 m^3, the amount of fuel left in the tank is ....

   (a) 15 litres   (b) 30 litres

   (c) 45 litres   (d) 90 litres

WORK

The two factors which govern the amount of work done are

....

Work done is defined as ....

....

....

Force is expressed in ....

Distance is expressed in ....

These two values would produce ....

But the unit used to measure work is the ....

and one .... = one ....

Consider the situation below.

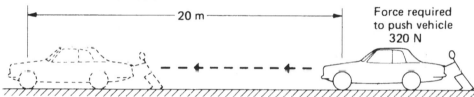

Work done by the man in moving the vehicle will be: 320 × 20 = 6400 N m

This is ....

Problems

1. A hydraulic lift exerts a force of 7000 N to raise a vehicle to a height of 2 m. Calculate the work done.

   ....

   ....

   ....

   ....

2. If the force required to tow a vehicle is 800 N how much work is done if the vehicle is towed for one kilometre?

   ....

   ....

   ....

   ....

3. Find the work done when an engine having a mass of 200 kg is lifted a vertical distance of 1.5 m.

   ....

   ....

   ....

   ....

   ....

   ....

5. The work done in dragging a metal box a distance of 6 m along a workshop floor is 726 N m. Calculate the force required to pull the box.

   ....

   ....

   ....

   ....

   ....

7. The work done in propelling a car 300 m was 72 kJ. What was the car's total resistance to motion?

   ....

   ....

   ....

   ....

4. Calculate the work done to drive a car 250 m if there is a rolling resistance of 50 N and a constant gradient resistance of 150 N.

   ....

   ....

   ....

   ....

6. The work done in raising the front end of a vehicle was 320 N m. If the force exerted by the lifting jack is 3200 N, to what height was the vehicle raised?

   ....

   ....

   ....

   ....

8. The work done by a press to push a bearing on to a shaft was 200 J. If the applied force was 10 kN, how far was the bearing pushed on to the shaft?

   ....

   ....

   ....

   ....

TORQUE

When a spanner is placed on a bolt and an effort is made to turn the bolt, the applied force is said to have created a turning moment or torque.

Define what is meant by torque ....

....

....

To calculate torque the following formula is used:

Torque =

Where the force is usually expressed in .... and the radius in ....

The SI unit for torque is ....

Complete the following exercise:

← 0.3 m → ↓125 N Torque =................... .... (a)	← 0.2 m → ↓125 N Torque=................... .... (b)
← 250 mm → ↓200 N Torque =................... .... (c)	← 150 mm → ↓0.3 kN Torque =................... .... (d)
Wheelbrace Crank radius 130 mm 25 N Torque =................... .... (e)	Tap wrench 12 N↓ 12 N↑ ← 0.3 m → Torque =................... .... (f)

ENERGY

Energy is obtained in many forms. It cannot be created or destroyed. It can only be converted from one form to another.
List the four basic forms of energy found in a car.

1. ....   2. ....   3. ....   4. ....

The engine converts the chemical energy contained in the fuel to mechanical energy which is used to propel the vehicle.

How does this energy conversion take place?

....

....

....

....

....

The battery-charging system uses the four basic forms of energy. Explain how the energy is converted from one form to another.

....

....

....

....

....

....

....

....

When solving simple mechanical and heating problems:

Energy can be defined as ....

The SI unit of energy is the ....

PRINCIPLE OR THEOREM OF MOMENTS

Using a simple beam (for example, a metre rule) which can be pivoted on a stand as shown below, apply forces of different value to either side of the pivot. Position the forces by the use of 'cord loops' to obtain a state of balance, as shown in the diagram.

NOTE:

Sum of the forces horizontally $\Sigma\ F_x$ must equal   0
Sum of the forces vertically $\Sigma\ F_v$ must equal   0
Sum of the moments both sides $\Sigma\ M$ must equal   0

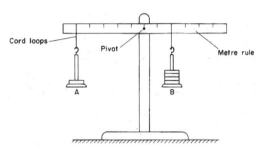

Add the information as required below:

force *A* = ...   force *B* = ...

perpendicular distance from pivot force A = ...

perpendicular distance from pivot force B = ...

Each force will produce a moment about the pivot

that is, a moment = ...

force *B* will produce a clockwise moment = ...

force *A* will produce an .... =

The product of both these moments is ....

It can be seen therefore that to obtain a state of balance (equilibrium)

....

There are considered to be three types of levers, all using the principle of moments in their operation.
The lever may be considered to be the simplest form of machine.

Type 1 has a force at one end while the resisting load is at the other.

The fulcrum being somewhere along the lever.

Type 2 ....
....
....
....

Type 3 ....
....
....
....
....
....
....

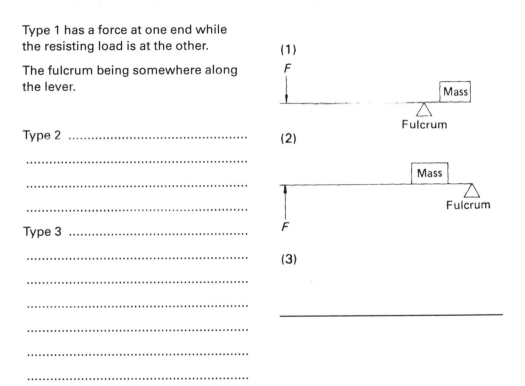

Show by simple sketches one motor vehicle or garage application of each of the types of levers shown above.

TOOLS USING LEVER PRINCIPLE

Many garage tools are based on the principle of levers. Study the illustrations below, name each one, indicate the position of the fulcrum and show the direction of movement of the effort and the load.

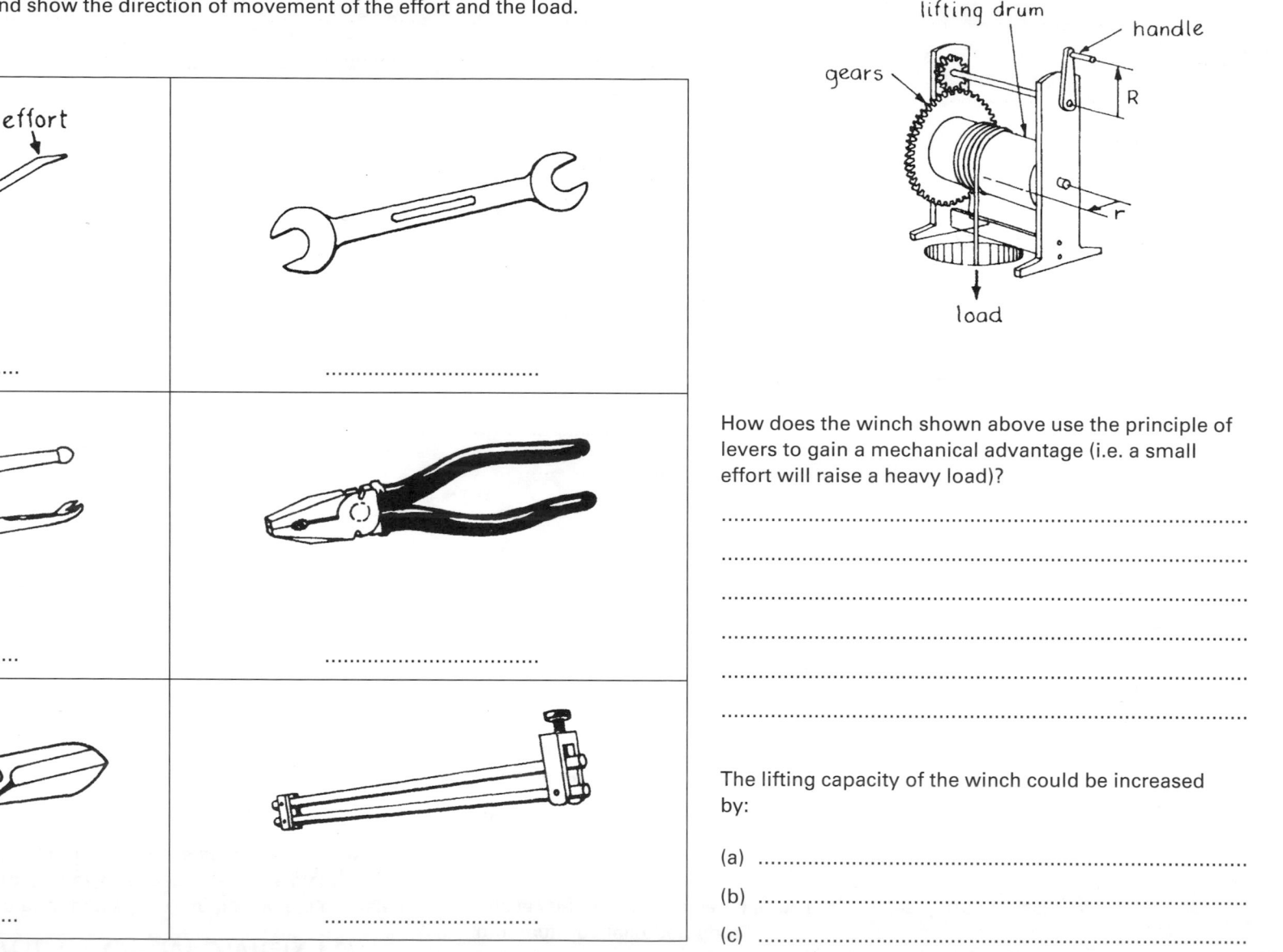

effort

load

pivot

....

....

....

....

....

....

WINCH

lifting drum

handle

gears

R

r

load

How does the winch shown above use the principle of levers to gain a mechanical advantage (i.e. a small effort will raise a heavy load)?

....

....

....

....

....

....

The lifting capacity of the winch could be increased by:

(a) ....

(b) ....

(c) ....

Problems

1. Determine *F* needed to balance the beam.

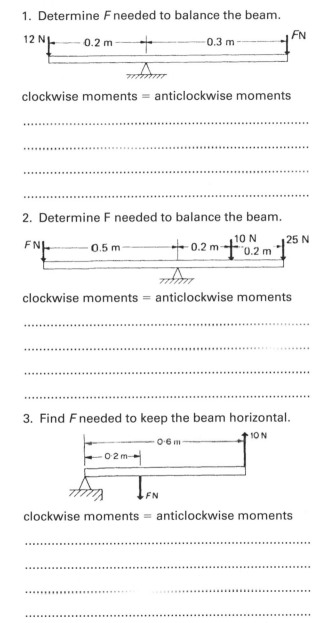

12 N ┃———0.2 m———┃———0.3 m———┃ *F*N

clockwise moments = anticlockwise moments

....

....

....

....

2. Determine F needed to balance the beam.

*F*N ┃———0.5 m———┃—0.2 m—┃ 10 N ┃ 25 N
                              0.2 m

clockwise moments = anticlockwise moments

....

....

....

....

3. Find *F* needed to keep the beam horizontal.

┃———0·6 m———┃↑ 10 N
┃—0·2 m—┃
↓ *F*N

clockwise moments = anticlockwise moments

....

....

....

....

4. Determine the force *F*.

Handbrake lever

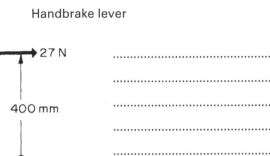

→ 27 N

400 mm

75 mm
F →

....
....
....
....
....
....
....

5. Determine the force *F*.

Brake pedal

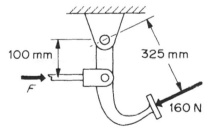

100 mm   325 mm

F

160 N

....
....
....
....
....

6. Complete the table below by adding values that will maintain equilibrium in the lever shown for the situations indicated.

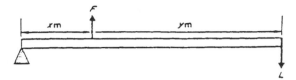

*x*m   *F*   *y*m
                          L

L (N)	*F* (N)	*y* (m)	*x* (m)
40		0.75	0.25
	150	1	0.25
8	40		0.5
100	1000		0.6

....
....
....
....

7. Determine F needed to maintain equilibrium.

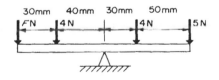

30mm   40mm   30mm   50mm
*F*N   4 N   4 N   5 N

....
....
....
....